The Unwanted Companion

A True Story

By Patrick J. Murray: Pioneer

ISBN: 978-1-326-26951-7

Published in London, 2013

Introduction

The gradual evolution of the world, towards being a better place for all its inhabitants, has already encompassed many things: from all the major inventions, to the development of all facets of medicine, to scientific discoveries and advances in science and technology, and to also confronting or eradicating some of the worst types of evils from the world. We can all see and measure this progress; by seeing the world of only a century ago to the world of today. But along this path, and as this evolutionary progress continues, there will be more new discoveries, and things we will be enlightened to; the answers to more secrets, and mysteries, will be revealed to us.

"My story is about one of them."

Contents

Chapter One

The Beginnings

I felt like the end was near, and felt close to death; like something was closing in on me. But strangely enough, there was a part of me that didn't care either. I was tired of fighting and battling. I'd had enough and was at the end of my tether. I had battled and adapted to all the other health problems, but this one seemed out of my control. There was nothing more I could do. I felt like my upper body was now collapsing, and at some point, and very soon, something had to give.

I'd been thinking about what I'd gone through, and was still going through. I had persevered and persevered, and now felt I could take no more. And in this deepening despair, I had quietly referred to God, saying, I wish you would just hurry up and take me out of this hell, I've had enough, I can't take anymore.

The will to try and get better; if Olympic medals were given for trying and sheer perseverance, I would have won gold. I had thought about what I'd been through over these last number of years and it had been a voyage through hell. And with the current situation, I saw no way out. It was now out of my control. I'd been through every emotion in the book, over and over and back again, and had had enough. I was tired and drained and worn down by it all. I felt like I had been on an endurance test.

But it was one morning, during this second week of August, during this very bleak escalation and worsening crisis that I had felt that something was coming to a head. And on this particular morning and hot summer's day that I had left my flat around mid-day to go to the hospital, because of an even severer development of additional symptoms, and had walked along Tufnell Park Road up towards Tufnell Park; which is about two thirds of a mile walk, and slightly uphill. It was a glorious hot summer's day, the height of summer and as I slowly walked along the road, tired, in pain, and feeling faint and light-headed, and with the added pressure of the hot sun beaming down on me, and on gradually reaching the steepest part of the road, about a hundred feet from the end, a real feeling of exhaustion came over me. I felt terrible and weak. I walked a bit further; and now, feeling fainter and even weaker, and suddenly feeling as though I was about to lose control of myself, it was at this moment that I fell forward, about to collapse and fall flat on my face. And too weak and powerless to stop myself from falling, and not even trying to fight it, I just thought to myself, 'Fuck it,' and felt like I was about to die. And as I fell forward, accompanied by a sense of losing consciousness, about to hit the ground, all of a sudden a powerful swirling mass of energy surrounded and engulfed me, stopped my falling, propped me back up, took control of me, and marched me up the hill. And as it did so, this swirling mass of energy seemed to be pumping into every part of my body; literally enmeshing me in it; swirling, pumping, surging into my body, my legs and up through my arms and swirling all around me and all over me, it was like a mini whirlwind, and continued until I'd been brought to the top of the road, then gradually tapered off, and seemed to disappear.

Bemused by what had just happened, and feeling a bit better, I paused, and stood there momentarily; then I crossed the road. And as I walked, I again felt the presence of this energy still swirling about me. But now much more gentle; and nowhere near as powerful. I walked a bit further, and sat down at a bus stop; and just sat there, not stunned, or amazed, but just very calm, tranquil, and feeling numb: knowing that something had just helped me, and something that was obviously otherworldly. And gradually realising that something must have been looking out for me, and may have saved me from death: or at least falling flat on my face and serious injury.

I sat there for a while, taking in what I had just experienced; and in the normal scheme of things I suppose I should have been amazed. But at this very moment, I wasn't. Experiencing something this amazing when you feel close to death doesn't have the same sense of awe as it would under normal circumstances, and would be safe to say that this would be the same probably for most people. But it was here, right here, where my incredible story began. But before I continue, I will tell you about myself, and tell the story that led up to this.

I came into this world on the 10th of September 1965. Born at 1.21 a.m. in Saint Asaph, North Wales, to a Welsh mother and English father; my father being from Liverpool. I was the youngest of five children. I had two brothers and two sisters. Between the eldest sibling and myself there was an eight year age difference; this was my brother Danny. He had been my father's first child from a previous marriage. My other siblings were close in age, all being born one year after each other with me being the last; closest to me in age was my brother Peter, one year older, and my sisters two and three years older.

My parents had met in 1960. My father, who for several years had been a soldier in the army and had been stationed at various army bases, some of them abroad, had been sent to Kinmel Camp army barracks in Bodelwyden North Wales, where the main focal point of activity for a day or night out for the soldiers was the resort and coastal town Rhyl, which was seven miles away. This was where my mother lived and was from, and where they had met. By the time I had arrived the family was living two miles from Rhyl, in a detached rented house in Kinmel Bay. By this time my father had left the army and had become involved in construction.

Although I was very young, my earliest memories are from the time that we lived in this house. And by the time we moved from here over two years later, and although barely two and a half years old, I still have some clear memories of being there with my brothers and sisters, and remember across the road from the house there was a small field with several horses.

By now we'd moved two miles away to Towyn. We'd been allocated a council house, which we lived in for four years; during which time my parents had been able to get a mortgage, and in 1971 we'd moved into our own house. It was a 3 bed semi-detached, was brand new, in a newly built street, and was also in Towyn, barely ten minutes walk from the council house.

Towyn is also on the coast, and had grown out of what originally was more or less a village; and we were just three quarters of a mile from the sea. This is where I would grow up!

The North Wales coast is tranquil and picturesque, mountainous and rural, with many scenic parts both stunning and breathtaking. Rich in history, littered with castles, it's not surprising that North Wales has one of the largest concentration of caravans and caravan sites in Europe: a good number of these sites in Towyn.

A great many of the people who would come here for a holiday were from the midlands, and the North of England. And with the seaside resort Rhyl in the vicinity, with its long promenade, funfair, amusement arcades, sandy beaches, bars and nightclubs, nearby Towyn was a good place to stay.

Towyn itself had many of these same attractions, but on a smaller scale. And some of the caravan camps had their own clubs and amusements. Towyn also had a smaller funfair. Growing up here was good. The autumn and winter months would be quiet, everything would close down. And during the holiday seasons and long summers, it would be brought back to life by the tens of thousands of people who came here for a holiday. Some of my fondest memories of childhood were during the very hot summers of the 1970s, when people came here in droves. It was like this every summer up until the mid 1980s, when the numbers began to decline. This was largely due to the increasing popularity of holidays abroad, mainly Spain, the Costa del sol etc, with its cheap package holidays and guaranteed weather. As of today the numbers coming here for a holiday is much less than it was.

The primary school we attended was catholic, and was in Rhyl, and which we were taken to and brought back by bus.

School sports day was a big occasion, this was something I enjoyed. I won the flat race (100 metres) almost every year, and sometimes won or came close in some of the other more themed type races too: the sack race, egg and spoon race, obstacle race etc. It was always scheduled two weeks before we broke up for the summer holidays, and cash prizes were given for first, second and third place.

I was also in the football team for my year. I was the goalkeeper. Our team was very good and one year we won a football trophy. This was for an annual competition between all the primary schools in a specific North Wales region, within the 9 – 11 age range. I missed out on the final two games that won this trophy; I had been dropped from the team due to misbehaviour. I had got in trouble from time to time and had been given the cane as punishment on numerous occasions, mainly for fighting, fist fights, etc. However on this occasion, I was totally innocent of any wrongdoing, but due to my reputation I was not given the benefit of the doubt. Looking back, I felt that the punishment was excessive, as at the time I had been making a conscious effort to behave myself. I clearly remember the football team going around all the classrooms showing off the trophy and then having an official team photograph taken with it. The photograph was hung along with other football photographs from previous decades, in the official school entrance. I sometimes looked at it and felt hard done by, thinking I should have been in that.

School was school, our family had an excellent attendance record, so much so that the headmaster had made a point of mentioning it; wind, hail or snow, we were there. We were very healthy children, rarely were we ill.

The school itself was well run and well disciplined, with very good teachers and an emphasis on religion and Christian teachings and celebrations.

One summer I got hit by a car and ended up with a leg broken in three places. My whole left leg up to the top of the thigh was put in a plaster cast, which I had on for six months. This happened during the summer holidays. I was nine years old at the time. It was my fault. I'd come out of an amusement arcade, and in crossing a main road had ran behind a passing car not realizing there was another one coming up at its left side: this is the one that hit me. I was fortunate that I only had a broken leg and had gone over the bonnet, and not under the car.

During the holiday season our mother worked in a popular fish and chip shop in Towyn; which also had a traditional restaurant attached to it. And during the school summer holidays almost every Sunday she would pay for me and my brother Peter to have a meal in the restaurant. In those days it was traditional to have a roast dinner on a Sunday, and that's what we always had, and came with a choice of dessert. And the restaurant had a nice decor and atmosphere, with pleasant music. Whenever I hear songs by the band 'The Carpenters,' it reminds me of some of these long hot summers of the 1970s, and of being in that restaurant.

Almost every Sunday up until the age of eleven we also attended church. But as we went on to high school, this gradually came to an end.

By the time I was thirteen and a half, I had been in high school only one and a half years and had been expelled. This was due to an incident that involved me and my brother Peter, and resulted in both of us being expelled together. The incident occurring after lunch break, when all the pupils would stand in line according to their year and would wait for further instruction, to be told, one line at a time to go on to their class. There would always be a few teachers on line duty. On this day there was a particular teacher; he taught maths, and had a reputation for being very strict and also a bit eccentric. He would often throw chalk at pupils in his class if they weren't paying attention. Also, he had a way of talking that when he ended a sentence or something he said, he would make a lingering noise that sounded similar to the noise a pig makes. On this particular day, as he walked in between the lines of pupils, some of them made pig like grunting noises when his back was turned. This would wind him up. There was quite a few of the pupils making these noises. And as he walked along the line I was standing in, someone in the opposite line made the noise, he looked around trying to see who it was, as he did so a lad standing behind me made the same noise. The teacher then immediately turned round then marched over towards me very fast, and thinking it was me punched me in the stomach. He hit me hard, it winded me. He then proceeded to slap and push me around, shouting at me as he did so. As he was doing this I was doubled over trying to get my breath. My brother who was standing in the next line ran over and shouted at the teacher, 'Leave him alone you bastard; followed up with a good swinging punch to the teacher's head. More punches and 'kicks' followed. A scuffle then ensued between the teacher and my brother. By this time I'd got my breath back and was very angry. I then turned on the teacher too, lashing out at him with kicks and punches which were full of venom. It was soon broken up, as all the other teachers soon came over. Me and my brother were then taken to the headmaster's office and made to stand outside, and wait. And after waiting for a while, the headmaster who by now had been informed of the incident called us into his office, and not really wanting to hear anything we had to say, simply informed us that we were being expelled.

I remember the day well, it was a Friday afternoon. The following Monday our father took us back to school. We stood outside the headmaster's office while he went inside to talk to him, and try to get us reinstated; the headmaster refused, and that was that.

We had been expelled for something caused by someone else. Although in reality this expulsion was the result and culmination of previous incidents of trouble that me and my brother had been involved in: fights as well as other things. And I had been given the cane here too on a few occasions, one time six of the best (three strokes of the cane on each hand) and this was the last straw. As the headmaster had on a previous occasion threatened to expel us.

I had always had respect for the teachers. It was outside of the classroom that I would get into trouble. And doing things to wind a teacher up wasn't something I would do. Even at that age I was more grown up, and had a bit more about me than that. And doing things like making silly noises just wasn't me. And it wasn't that I looked for trouble either but sometimes trouble seemed to find me.

This happened a few weeks before the summer school holidays began. By the time the holidays were over and a new term had begun, the schooling board had decided what to do with us. For the following year I was given what was termed home tuition. A taxi would pick me up from home and take me to a primary school in Rhyl, where there was a small room set aside for me; and where I would receive tuition from a teacher. This was for three hours a day, 9 a.m. till 12. The taxi would then pick me up and take me home. My brother was for a while sent to a place in Rhyl called the adjustment centre. And during his last school year also received home tuition, with the difference being, the teacher would come to our house for three hours. I didn't particularly like the home tuition. Although the emphasis was on continuing my education, I felt isolated, and it seemed that there was an element of punishment about it.

Still, life wasn't so bad. I was now 14, and life was becoming more interesting. I had worked during the summer school holidays on the funfair in Towyn, next to the beach. I had been given a job as a ride attendant. I operated a small ride for children and looked after the trampolines, for which I collected the tickets and timed and organized. And by the time the summer season was over I'd saved a lot of what I'd earned.

I had also been since the age of 11, a big fan of 1950s rock n roll, and by now, had adopted the combed back Teddy boy hairstyle; as well as buying certain types of clothes and shoes to fit in with the image. That's me on the front cover aged 14/15, sat on my bicycle. Some of the first songs I had heard from this era were by Jerry Lee Lewis, Little Richard, Gene Vincent, and the Big Bopper. And I enjoyed discovering and buying records from this era and these artists. I even saw Jerry Lee Lewis in concert. He toured Britain in 1979, and played at the nearby Deeside Leisure Centre. During school terms I'd also started attending the local youth club, which was one evening a week, and sometimes took my records there to play them. I also liked playing table tennis and became very good at it, one year winning the youth club tournament.

It was also during this period, from the age of 11 to 14 that I had got into trouble with the police. And over a two and a half year period had appeared in court four times, first at age eleven. In them days a juvenile would only have to get into trouble a few times and could end up doing some time in Borstal. I remember at age 14 two brothers I knew aged 14 and 15, being sent to Borstal for two years each. And I remember during my last appearance in court around this time, the Judge saying that if I got in trouble again I would be sentenced to a stint in Borstal; and because of this, made a real effort to stay out of trouble.

For my remaining two years of school, I was given a place in a school for problem pupils, who, like me, had been expelled from other schools. This school numbered probably no more than fifty pupils. And class sizes numbered no more than ten. It was part residential, with half of the pupils resident. I was not one of these; I was picked up from home and taken home by taxi.

The school itself was located outside of Hollywell, which was about twenty miles from Towyn. It was on a high point, almost a hill, and surrounded by miles of barren moorland, with a winding road cut through it. And although a long drive, it was pleasant. By this age I was smoking cigarettes, and having a smoke often enhanced this pleasant drive. Most of the time I had my own cigarettes, but when I didn't I would sometimes ask the taxi drivers for one. I had some good conversations with the taxi drivers; one in particular was very interesting. He told me some great stories of when he was a soldier fighting in the Second World War. From battles he was involved in, to his part in the liberation of a concentration camp.

The school was easy-going. There was no real pressure put on pupils to learn anything. It was pretty much basic education with a bit of history thrown in. This suited me, I wasn't bothered. As my aspirations at this point were more of a musical nature. And I could already read and write, and was good at maths. And the level of education that was taught I already knew. The history that was taught was varied, and I enjoyed history, also woodwork, which I was good at; and occasionally there was some sport, generally rugby; and the occasional 4 mile cross country marathon through the surrounding mountains, which the first and only time I participated in I won. Also, I had for some time wanted to learn to play the guitar, and noticing that there were two teachers who played guitars when hymns were sung during morning assembly, I enquired about having lessons, and started having lessons soon after. The first song I learned to play was the song 'Day Tripper' by the Beatles. I pretty much learned to play the guitar on Beatles songs, and became quite good.

By the time I was 15 or 16, I was trying to get involved with music, attempting on a few occasions to join a band, but not really finding the right one; and was by now getting more influenced by music from the 1960s. I'd even written some of my own songs, and later tried to form a band myself. Of what I did form nothing came of it.

Up until the age of sixteen I had continued to work during the summer holidays on the fairground; the last season I worked running the go-carts, and by now had also left school. I had decided during my fourth season working on the Fairground that that would be my last. It was ok while I was in school, but felt I needed to move on: to do what I wasn't sure! In the early 1980s much of the country was in recession with three and a half million unemployed, and being a school leaver, looking for a job was hard going. I'd applied for a number of jobs, and it seemed like there were dozens if not more applying for each job, and they all seemed to want qualifications of some sort for every job no matter how mediocre the job. And the pay for many of the jobs was equally depressing. And the fact that I didn't have any qualifications didn't help either. I could go to college and learn a trade, or the other options available were: Youth Training Schemes. And after several months of trying to find something, without any luck, and because of the boredom and general gloom in the air, I had even come close to joining the army: something which had been arranged by my father. And only due to a mix up with times and missing a lift had later changed my mind. I had tried out the Army Cadets several years earlier, but hadn't stuck around long enough to acquire the uniform. The Youth Training Schemes! This was something that the government had introduced a couple of years earlier, for 16 to 19 year olds. The idea was that an employer would, for six months, employ someone, and during this period the government would pay the employee £25 a week. Even in 1982 it wasn't a lot of money but it was I suppose a way of creating jobs, because when the six months came to an end some of them were kept on. Although from what I gathered it wasn't a lot. Most employers saw it as a way of getting six months free labour, and had no intention of giving you a job at the end of the six months. The plus side of it: work experience and learning!

I was by now 17. And after finishing that last summer on the Fair, and spending those following winter months trying to find a job – resulting in nothing, I had now decided in March of the New Year to enquire and take up the offer of the first of two Youth Training Schemes, both of them on holiday camps in Towyn, the first lasting only five weeks, the result of a falling out with the boss. And the second, I started soon after and lasted the whole six months. This was for a company that owned three big holiday camps, two of which were in Towyn. These were proper holiday camps, with their own amusements, night clubs, shops etc. And a plus side was that they topped up your £25 with an extra tenner – which made a big difference; £35 a week wasn't bad. And during the time that I had started there, with the main holiday season near, the main work that was being done, was getting the camps ready: maintenance and repainting etc. And although mainly working on one camp, I also alternated between the two. And when the holiday season took off the worked changed, and the main tasks were attending to any problems people had in their caravans or chalets, and fitting new gas bottles, cutting the grass, emptying the bins etc; and all in all the job was enjoyable.

I had also passed my driving test during this time, and had acquired a 1972 Ford Escort for £50. Although it had a few mechanical problems, it was soon sorted out. Added with a new paint job, I soon had a good car. Having a car for the first time was great. And I was also given an additional job as a glass collector in the night club on one of the camps. And this, for many reasons was a good place to be also.

I didn't expect to be kept on when the Youth Training Scheme came to an end and knew I wouldn't be, as it would then be out of season. And when the holiday season ended and the six months came to an end in October, I moved on. It had been a good six months and a memorable summer, and I had been treated well, so I had no complaints. Now 18 years old: my mother by this time had become a care worker. And my father who had been involved in the building industry for many years had a lot of work on. So I ended up working for him. I, along with my brothers had worked for him in the past, usually during school breaks; and sometimes taking time off school to help out, which I didn't mind, as the money he paid me was always welcome. The work had varied over the years, from steel fixing (the steel reinforcing framework inside concrete structures and foundations) which had been his main job for many years, to general building work: renovating properties, building extensions, garages, walls, a whole mixture of this type of work, which was often interesting and also kept you physically fit. He was also now a qualified Clerk of Works. This the job of making sure all aspects of building and construction work were checked and done properly, to the standard required; and had already been overseeing large scale building projects in this role. And was much sort after for his knowledge and expertise in this field. And over the years I myself had learnt a lot about building, and also became very good at carpentry and joinery.

Having a car also added a new dimension to my life. I had long been interested in metal detecting. In the newspapers from time to time you would read about a metal detector finding something really valuable. And this and the sense of adventure had encouraged me to get one. And now, with my own metal detector, I, along with a friend, often went detecting on the weekends; usually on or near historical locations, hoping that we might stumble on a horde of Roman coins, or other kinds of treasure. This never happened! But nevertheless it was an enjoyable hobby, with the chance of the unexpected, and one that took us on many pleasant drives throughout North Wales.

Chapter Two

On My Travels

By this time I had also become fascinated with travelling abroad, and during the spring of the following year, along with a school friend, travelled by coach and boat to the Spanish island Ibiza, for what we hoped to be a working type holiday, which only lasted three weeks due to us being too early in the season and running out of money. I had been in a similar situation 15 months earlier. A month before starting the Youth Training Scheme, I had, along with two acquaintances, gone to Belgium and Holland hoping to find jobs, first visiting Belgium then Holland, to Amsterdam and Rotterdam. And after running out of money we had ended up being deported. We had been waiting in the Euro Port in Rotterdam for money to arrive by telex, the last of my savings, which would have enabled us to stay a bit longer. And which we were told would take three days, and had spent two nights sleeping in the luggage carriages outside. And on the third morning had attracted the attention of security people, who asked what we were doing there, then took us to the police station and an hour later informed us we were being deported. And the next day we were back in Britain. The official reason we were given for being deported was vagrancy -- which made us all laugh. And although not having the desired outcome, it had been the first time I had been out of the country and had been a good adventure, with plenty of laughs along the way.

But it wasn't long before I was off abroad again. Now a year since the visit to Ibiza, and this time going on my own, I had decided to go to Munich, and had arrived there the first week of May 1985, travelling by coach. I was now 19. And recall when the coach journey was nearing its destination, how the prospect of then having to get off had seemed slightly daunting. As well as being alone in a foreign country, I didn't speak the language, it being night-time; and I remember thinking that I wished the coach journey would last a few more hours. A change, the sense of adventure, and the idea of living in a city and a foreign country for a while, had brought me here. And Germany then was booming; at the time having the strongest economy in Europe. And after initially staying in a cheap hotel (cheap by Munich standards) for a week, then buying a sleeping bag and sleeping rough for several days to conserve money, while looking for somewhere cheaper to stay, and also see what work was available; I stayed with some German conscientious objectors for a few weeks, and they had helped me get an evening job as a dishwasher in a restaurant in an upmarket area of Munich. I had met them through asking someone sat opposite me in the main train station for a light, who had turned out to be British. I then stayed in a tent in a large campsite next to a river on the outskirts of Munich, and with the summer weather approaching it was a good place to be, it having all the necessary facilities, as well as a social club, and had soon got to know some of the people in the other tents around mine. And during the following months, as well as working I often went sightseeing; once visiting the former concentration camp Dachau.

I worked the dishwashing job for three months. The money had been good, fifty marks a night; which worked out at about £90 a week, plus a meal. If I'd been earning that in Britain I would have been doing very well. Then I later worked as a glass washer at the Oktoberfest; the famous beer festival, lasting three weeks. Then sweeping leaves for a couple of weeks. Following this there was a few months work in a printing factory stacking packages on pallets. By this time I had left the campsite due to it closing for the winter. And I, along with someone I had become friends with on the campsite, was staying in a shelter for down and outs. This was Gerry, from Dublin, who was the same age as me, and had come to Munich after being sacked from a student type job in another part of Germany. And through a girl he had been going out with, had been instrumental in getting me and himself the jobs in the factory and sweeping leaves

We had often talked about travelling the world and had later made a plan to travel overland to Australia, visiting many of the countries along the way, with Singapore being the last – then by boat to Australia, where we hoped to work for six months. And we were hoping to start this in a few months time. But first, in the New Year, in January, we travelled to another German city called Manheim, and then on to a small town not far from there; to a clinic which tested new medications on people and were testing a high blood pressure pill, and needed a number of volunteers. Gerry had found out about this and had contacted them, and we were told to come for a medical examination, which we passed, and were given a place on the course, plus accommodation. It was a six week course of tablets, and the pay was 5000 marks, which was very good money. And during our time there we planned for our overland journey to Australia. And after leaving the clinic, and with no ill effects we set about putting our plans in motion. First, travelling to Bonn the then capital city, to visit the necessary embassies for visas; and where we stayed in another shelter for down and outs. And over the week managed to get visas for most of the countries, but had to wait for the Australian one.

One evening in the shelter a man attacked me. I'd gone into the toilets to get some toilet paper to blow my nose. As I came out of a cubicle a man then went into it. He then turned around and started shouting at me, pointing to the toilet seat: someone had obviously urinated all over it -- and he was blaming me. I hadn't even used the toilet. I had learned some German by this time but it wasn't good enough to explain the situation, and as he was becoming more aggressive I got frustrated trying to, and walked out. Ten minutes later he came into the dormitory, shouting, and looking over at me. I was stood up and talking to Gerry, who was sat on his bed on the top bunk. He continued shouting and looking at me and pointing. And now angered by him, I then shouted some abuse back at him in English. He then took affront to this, becoming even angrier, and charged down the aisle between the beds, now running towards me. Now expecting a fight I turned sideways. He then kicked me hard in the leg. He followed this with punches. But I had been in many fights before, and fighting back was second nature. And after absorbing his initial blows I pushed him hard with both hands, unbalancing him. Then repeatedly punched and kicked him as he tried to regain his balance. And although doing his best to fight back I continued to unbalance him, and caught him with some first class punches to his face and body, and eventually knocking him to his knees, where I continued to rain down blows on him: just to make sure he didn't get up again. And it was at this point someone intervened on his behalf.

He had charged at me with the intension of doing something similar to me, and had come unstuck; all the result of a misunderstanding.

It was a couple of days later at the Australian Embassy that we were informed that we had been refused visas; citing, that they didn't believe we were just going for a holiday. We didn't fit the criteria for work visas so had applied for holiday visas instead, which left our plans in ruins. But walking past a travel agent's a week or so later we had seen a notice for a cheap flight to America: $100 for a return ticket to San Francisco, California. This was very cheap. And although we had not planned to go to America yet, it seemed like a good idea. We discussed it, and the next day had decided we would travel to America instead. And two weeks later, the second week of March 1986, we were on a flight to San Francisco.

Germany had been good. And while there I had managed to save a lot of money. I had been there almost a year. And it was strange leaving. And touching down in America was equally strange. We stayed in San Francisco for three weeks. Staying in a house affiliated with a youth hostel, whose owner had converted some of the rooms, putting bunk beds in and charging mostly foreign students $10 a night. I had even been given a job for a couple of weeks while there, working for a self-employed carpenter, who had been staying in the house and had got talking to Gerry, who had mentioned that I was quite experienced at carpentry. I had had similar good fortune in Germany; just being in the right place at the right time.

We moved on next to the seaside town, Santa Cruz, which was somewhere between Los Angeles and San Francisco, staying for a week. Before we had left San Francisco we had seen many of the sights and places of interest, including a tour of the former island prison, Alcatraz, now a museum. And after a week in Santa Cruz, we had boarded a coach for Los Angeles. And I remember as we were arriving there, seeing the famous Hollywood sign; and was surprised, as I hadn't known at the time that it was in Los Angeles. And after initially staying in a youth hostel at Venice Beach, we later found accommodation close by in the form of a studio type room in the rear garden of a house. We had now decided we would stay in Los Angeles for a few months and try and find jobs. And while here plan where we would go next. And in the meantime we were seeing the sights and enjoying the nightlife.

We were later offered jobs by the person who had rented us the studio: repainting address numbers on kerbs. Something which every house in Los Angeles has on its kerb; usually a white square background with stencilled on black numbers, next to the driveway, and which needed to be repainted every year or so. And the way this was generally done was that the street would be leafleted to let the residents know they were coming round to repaint them. And the next day they would be painted. And after they were all done, you would knock on the doors and ask for a five dollar donation; and which out of the net sum we would get a percentage. And although initially I wasn't interested or enthusiastic --- the idea of having to knock on doors, and ask people for money didn't appeal to me at all. But Gerry along with six others started a week later. Then after a week I decided to give it a try also.

One evening around this time, while in a bar, I met a girl who I started to meet regularly: an American named Susan, who was two and a half years older than me, and from Road Island. And who had arrived in Los Angeles four months earlier. And although I hadn't particularly liked the curb painting, for the time being I stuck with it.

Two months later me and Gerry left the studio and were staying in a cheap motel along with another of the curb painters, Scot. And had decided to paint curbs ourselves. But instead of all the extra work of painting the whole street, then asking for donations, which many people avoided giving you, we would leaflet them first then knock on doors to see who wanted them repainted, and at the same time giving a bit of a sales pitch, reminding them of the importance of having the numbers clearly visible; for emergency services, and other things etc. This way we would get the five dollars for each one.

The change of circumstances had come about as a result of the man who we rented from and were working for, being served with an eviction notice for non-payment of rent on the property that he was living in, and renting to us. And as well as him not paying his own rent, he was also avoiding paying the people who worked for him, which there were twelve of. He owed Gerry six hundred dollars and me two hundred, which I'd written off. Gerry somehow thought he would eventually be paid; he was dreaming! But things caught up with him. And one guy he'd owed money to for a while had attacked him with a thick length of hardwood, resulting in a large number of stitches to his head. He had pulled up in his car one evening just before we had left, and to our amusement with a blood-soaked towel wrapped round his head, and saying what had just happened and telling his wife to drive him to the hospital.

By now, I had been spending more and more time with Susan. And although me and Gerry had planned to travel America for the time being I wanted to stay put. But Gerry wanted to move on. And decided along with Scot to see other parts of America; going first to northern California to the Yosemite National Park. I had borrowed Susan's car and given them a lift to a campsite outside of Los Angeles, where they intended to travel from. It was sad to see him leave. Gerry had been a good friend. It was the last time I saw him in America.

Two months later, I celebrated my 21st birthday. And by this time I was living with Susan in an apartment we had rented together. A year later we were still together, and had moved to a less expensive apartment, located on the edge of Hollywood. There had been girls in the past, but Susan was my first serious relationship.

I had spent part of the year first working as part of a carpentry team on a construction site; and when that ended, decided to do the curb painting on my own. Then later I had hooked up with another of the curb painters I had initially worked with, named Dan, a musician, rock guitarist, from Michigan, who was trying to make it in the music business. Hard rock, heavy metal, was big at the time and was experiencing an explosion of bands of this type of music, and there were many nightclubs of this nature in and around Los Angeles. And a number of which were on the sunset strip, which I myself sometimes frequented, as I had increasingly grown to like this music too. We had our own transport, had made up a better and bigger leaflet, made our own hours and made quite a good living, working around most of Los Angeles and the surrounding suburbs; from some of the wealthiest areas to some of the poorest. And by now I quite liked the curb painting. And the ones we painted, they were neatly done and looked sharp. And as well as being my own boss, and making my own hours, I didn't have to deal with someone moaning at me for being five minutes late either.

I stayed here for the next couple of years. And during this time I had, along with Susan, visited her home town in Road Island, on the east coast of America. And while in that vicinity had also visited places of historical importance, like Plymouth Rock where the first pilgrims had landed, as well as the city of Boston. And had also been to many places of interest in and around Los Angeles, like Disneyland, Magic Mountain, Universal Studios etc. It was also during this time that Susan had started going to university. And during Christmas of 1987 my parents had come to see me.

A year later we had moved again, to a nicer area and a better apartment in West Los Angeles. And the following year of 1989 I had spent the summer in London, and had come back to L.A in October. And while I had been there Susan had visited me, and I had shown her some of the sights. I also took her to Wales, showed her where I was from, and took her to a castle. And by July of the following year I was back in London again, deciding to go back and work there for the summer, which I did, mostly working as a painter and decorator, and arriving back in Los Angeles a few weeks before Christmas; and along with Susan spent that Christmas in Mexico. First visiting the California Mexico border town Tijuana, and then driving to the small coastal town Rosarita, where we stayed at a quiet beachfront hotel. In its heyday, it had been popular with the American film stars of the 1950s. Rosarita reminded me of something I had seen in western movies -- more so a spaghetti western. And there was a sense of being transported back in time to this era. It was a relaxing pleasant few days, in another country, with a different way of life.

By now, although still doing the curb painting, I'd had enough of it. And in January of the New Year, now 1991, enrolled in a night class to do a real estate course, to get the licence to become an estate agent. The course itself was for one evening a week over a six month period, and required a lot of study. Susan, by now, had graduated from university with a degree and had been applying for jobs, and had been offered a position with Toyota at their USA headquarters located outside of Los Angeles, which she took and started in January, and was given a company car. And a few months later although still together we were now living apart. She had moved to a place closer to her job. And I had moved into a small apartment, in the complex where Dan lived in Lennox, not far from Los Angeles Airport. And in July Susan was in the process of moving to New York; after having been offered a position there by Toyota. And by August she was there. I went and stayed with her for ten days in September, spent my birthday there. 'Seen the sights'. It was an enjoyable time. And back in Los Angeles. After years of trying to make it in the music business Dan had decided to leave Los Angeles, and go and live in Las Vegas; and left in October. And although up until this point me and Susan had kept in touch by phone, September in New York, that was the last time I saw her. And by November we had parted and gone our separate ways.

Chapter Three

London

I had taken the exam for the real estate licence earlier that August, and failed. But planned to take it again, after restudying the areas I had failed on. In the meantime, I was still doing the curb painting, but now on my own. And now, with increasing competition, it was now harder to make a living. And by April of the following year, now 1992, one week before Easter, I was on a flight back to London.

Life over the last few months hadn't been good. Often feeling fed up and generally feeling down, I decided I would go back to London for a few months and work and continue to study for the real estate test while there. Plus I wanted to visit my family, as my father had been diagnosed with cancer of the oesophagus (the area that connects the throat to the stomach) and had recently had an operation, which, as a result of it being caught early enough before it had spread, was a success. And I had been in London for only a week when I got involved in some trouble. I had misplaced my passport. And thinking I'd lost it, had gone to the passport office to report it, and make an application for a new one. And in the confusion of the queuing system had got into an argument with a man over who was next in the queue, which escalated into a brawl, and culminated in me being arrested and charged with ABH (actual bodily harm). This happened on a Thursday; the next day being Good Friday, the start of the Easter bank holiday. And after being charged, was informed that I would be up in court the following Tuesday. And due to not having a permanent address was also informed I had been refused bail. And spent the four days in a cell at the police station, during which time I spoke to a solicitor. And on the Tuesday morning was taken to court where I was granted bail and given a date to appear in court, in six weeks time.

I was disappointed with the whole situation. And thinking about the scenario at the passport office I know my frame of mind hadn't helped me. And the fact I had recently started the process of giving up smoking, this had made me bad tempered. And had contributed to my behaviour, in the sense that in my usual frame of mind I would probably have behaved differently, or not as aggressively in the initial stages of the argument, which set the tone for the further escalation, and into what it became. And although I was somewhat to blame I was not fully guilty either, as I was further provoked after I had first tried to walk away. And also after a security guard had come over and I had gone and sat down, I was then threatened; and that's when I snapped.

I appeared in court six weeks later. And on the advice of a solicitor, reluctantly pleaded guilty to the charge. I was fined and put on probation for one year. The probation was a real blow, as it meant I could not go back to Los Angeles in a few months time as I had planned. And meant I would have to stay in Britain for the next year.

In the time before the court appearance, I had visited my family in Wales, spending two weeks there. And on returning to London set about trying to find a job, painting and decorating as I had before. I had heard there was now a recession in the building industry, as well as a recession in general; and after a month or so of trying to find a job was very much aware of this, as the lack of jobs were reflected in the sections of the main advertisers – there was very little advertised. And although I continued to look for another month, the fact that I had now received a probation order had made me rethink my whole situation. And not having much luck, I decided I would look for a job doing something else; like working in a bar or a pub, as some of them also offered accommodation, as I was presently staying in a bed and breakfast hotel. By now I had already started seeing a probation officer, initially once a week and after two months became every two weeks. And during this time I had been on several interviews for bar jobs, but had no luck. This pattern continued; life was miserable. And the last interview I went on had even resulted in another incident. I had arrived at a pub to be interviewed in the early afternoon, at 1pm, was told the landlord was in a meeting and would have to wait for a while. He showed up sometime after 2 o'clock, by which time I'd waited one and a half hours. And, for a while, I had waited patiently; but was now in a foul mood. He had further irritated me by being seemingly unconcerned, and saying nothing that would pass for an apology. He quickly looked through my application form and asked a few questions, then ended the interview by saying he would let me know. As we stood up he also said he had a further 20 people to interview yet, to which I responded, 'Well don't keep them waiting for an hour and a half or you'll never get through them.' He said, 'Well you wanted the job.' I said, 'Fuck your job,' and proceeded to leave. He quickly followed me, shouting, 'Get out of my pub,' and pushed me as he did so. I then turned around facing him, saying, 'Don't push me.' He said, 'Get out,' and shoved me again. Although a slob, he was quite a big feller and although I didn't look for trouble and was generally polite, I wouldn't be pushed around or picked on either and tended to have a short fuse when this occurred. And now incensed by him, I spat in his face, and head butted him. Punches were quickly exchanged, a fight developed. And it wasn't long before I felt myself being grabbed by two or three other men – customers, who eventually forced my arms behind my back? I was trying to break free from them, and in the process they had they managed to ram me head first into a slot machine. And by this time all three had got a firm grip on me; and one of them said: "You either calm down and leave, or we will call the police."

It had only been two months since the court appearance, and the thought of the police turning up quickly brought me to my senses. I thought that's the last thing I need, and calmly said, 'I'm leaving!' They let me go. I walked outside, relieved that they had given me a choice. I walked towards a roundabout and remember just leaning on a railing, and looking out into the distance, not really knowing what to do next. Since getting off the plane, it had been a miserable, unpredictable, and turbulent four months. The whole plan had been a disaster. And it was around this time that I considered leaving, skipping the probation, and going back to Los Angeles. But the trouble with doing so, meant, whenever I came back to Britain, if stopped by the police for something, I would be arrested. And would probably end up doing a prison sentence, as in many ways, probation is an alternative to a sentence.

I eventually decided I would stay put and see through the probation. But knew I would have to go back to Los Angeles temporarily to get my belongings, and clear up any business. Having borrowed some money for the flight from my father, I did this the second week of August, arriving back in Los Angeles and staying just under two weeks. And being back there had made me realize, that I did have no real reason for going back anyway; and certainly no longer any real enthusiasm for being an estate agent either. The main reason I had stayed there had been the relationship I was involved in, which no longer existed, and that was it really. And although there were some things I would miss, I finally said goodbye to Los Angeles, and that was that. In all reality, I think the judge may have done me a favour.

Not long after coming back it was my birthday. I was now 27. And back in Britain permanently. I had spent almost seven years living abroad. And in some ways it was nice to be back in my own country. The sense of belonging, that is lacking when living in a foreign country, had returned. I was home. But little did I know what was in store.

It was a month later during that September that I started a two year drama course, two evenings a week at the City Lit in Covent Garden. I had done an initial class earlier in July one half day a week lasting five weeks, which had qualified me for the two year course. I had done the class in July, initially to explore the idea of acting as a vocation, after I had seen it in a prospectus while looking for other career options --- after deciding to stay. And later decided on the two year coarse as a career move. I was also informed by my probation officer that the probation service ran a drama group one evening a week, which I also started attending in October.

It's strange that I had decided to go into acting, as over the years that I had lived in Los Angeles, from time to time people had said, because of my appearance, that I should become an actor; as I was in the right place for it. And at the time it hadn't appealed to me. And even during my last few months in Los Angeles, I'd gone out with a woman who worked in the Film Industry and had offered to get me a role playing one of the characters in a film that she was currently involved in, which was being filmed in a ski resort. And she had taken photos of me, had given me the script and offered to fly me out to Colorado where it was being filmed. I had looked over the script but didn't feel good about it all, or have any real desire to do it. And my frame of mind wasn't too good either; and around this time wasn't sure what I wanted to do anymore. So I declined the offer. But it had got me thinking about acting, and now all of a sudden it seemed like a good idea.

I had also now moved to another part of London. To a bed and breakfast hotel in Paddington, which one half of was occupied mostly by homeless people, and paid for by the state. And, as I was in this category, with no permanent home, and unemployed, I was given a room there. It was an improvement on the last place I had stayed in, as the best thing that could be said about that was that I had a roof over my head. Five months later, by February of '93 I had to leave the 2 year drama course. This due to what I was told was inadequate progress. I had been notified by mail that I could no longer continue on the course; citing also, clashes with other pupils as the reason – which had been exaggerated! It being more due to this drama teacher taking a dislike to me from day one; as by this time I had been making more progress and my confidence growing: some of the other pupils remarking how good I had been in some of the improvisations. And although I had felt angered, and had even later carefully planned some revenge; it was the knowing that I still had the other drama class to go to, that enabled me to eventually manage to rise above this and move on. He had narrowly escaped a trip to the hospital.

Life in general hadn't been much better. My room in the hotel had been broken into and my music system stolen; that was the only thing I had that was worth anything. And after informing the management, they moved me to the other side of the hotel to another room, which was nicer and had a television, and also a bathroom. This improved my quality of life. In fact it was worth having it stolen just for that. But it wasn't long after this that the hotel had a sudden influx of refugees from the conflict in Bosnia. Up until this point the place had been quite peaceful. But after a few weeks the whole hotel was full up with them, and it changed the atmosphere. There was always constant noise of some sort – loud music, banging, shouting, slamming doors, people running up and down the stairs, this kind of thing, and often until the early hours of the morning. After seeing the images on the T.V. of the conflict and the atrocities going on in Bosnia I had initially felt sorry for them, but after a month they were driving me round the bend – often being kept awake by them. And living there had become stressful. It's strange how circumstances can make a person go from feeling pity and sorry for people to despising them. After two months some of them had been given jobs in the hotel; cleaning, serving breakfast, and at the reception desk. I would go for breakfast in the morning and sometimes they would be playing Cossack type music and dancing to it -- at eight in the morning! One time I almost got into a fight with one of them for taking away my pot of tea without asking if I'd finished with it. He'd done it to me twice before and it wound me up, as the teapot contained enough for two cups of tea and he hadn't bothered to ask if I'd finished with it, which on this occasion I hadn't. It was a lack of manners. And this third time he did it I snapped, raising my voice, taking it back, and saying: 'Why don't you ask first if I've finished with it?' He leaned across the table towards me and saying in a raised voice, something in his native language; whatever it was it sounded nasty and threatening, and feeling threatened I quickly leapt up towards him and swung a punch at him – directed at his head. But the weight of the table pushing against the top of my legs as I leapt, threw me back into the chair and I missed. I then threw the table over to get at him, and before I did a lot of his people had quickly got in between us and diffused the situation. But he got the message. And it didn't happen again. In fact, after that he didn't bother clearing the table until I'd left.

Times were hard; I'd had very little money, and often went hungry and had sometimes resorted to walking the three miles to Selfridge's department store, on Oxford Street, to eat the samples in the food hall. The fortnightly money I got off the state, rarely lasted two weeks, and I was always scraping to get by. I had made many other attempts to find work, all unsuccessful; and had temporarily given up looking. It had been like this for almost a year. It seemed like I had a curse on me; I couldn't get a job doing anything.

It was around this time while looking through a London newspaper I had read an article about a magazine called the Big Issue, which was put together for the benefit of the homeless and destitute with the aim of giving these people a chance to have an income by it being sold exclusively by them. The article had been about a woman who had become destitute and how her selling the Big Issue had given her an income and a lifeline which eventually enabled her to get her life back together. The magazine had originated in America and had been established in Britain only 18 months. It was a good discovery. I bought a copy, and after finding out that they were based in Victoria went there and after being told the rules, photographed, and being given an identity tag and the first fifty magazines on credit, a couple of days later I was selling it. I had found a pitch by the tube station in Knightsbridge; the magazine sold for 80 pence of which 40 would be mine, and although initially I thought by having to sell the Big Issue I'd hit a new low, this thinking soon changed; the first day I made £24, and was very pleased. It had been the first money I'd earned in a year.

I continued to sell the Big Issue for another three months, usually 3 or 4 days a week. Some days were better than others, and some days were awful, although the most I ever made in any one day was probably no more than £25. But at least I was making something and life had improved as a result, it had helped me get out of a hole and I was grateful for that.

In May of '93 I moved out of the hotel, and also left the Big Issue behind with it. I had been given a room in a hostel in Islington, North London. It was for male ex prisoners and others like me on probation, who had no permanent home. The benefit of staying there was that after a year you would be nominated to a housing association, which meant I would eventually be given my own flat. I had heard about this organization at the probation drama group – which I was still going to. A person I had got talking to had put me in touch with them. Part of the criteria for being given a place there was that you had to stay out of trouble. The hostel contained about 25 rooms, and I had been allocated a room on the top floor, which I was pleased with. More than anything it was a relief to get out of the hotel.

I now felt things were moving in the right direction. And it wasn't long after this that my one year probation term had come to an end. I still continued attending the probation drama group and by the summer we had put together an improvised play which we performed at the 'Round Festival' in southern England. I'd also got a job as an industrial cleaner. The work was mainly in newly built properties, preparing them for being handed over after completion. Some of it was the hardest most tedious work I'd ever done, and for the lowest pay. But although a real shit job, it was a step up from selling the Big Issue. And as that September arrived, and the summer of '93 drew to a close, I was again working as a painter and decorator. And I'd also enrolled in a drama production group at a Westminster college, one evening a week, whose aim, as well as acting training, was to put on a production of plays. And as this progressed, I eventually left the probation drama group. I'd gained a lot from the group but felt it was time to move on. Strangely enough though, a couple of months earlier, a film production company had come to the probation group and had later auditioned a number of people in the group, including me, for parts in a drama documentary that they were making. And most of us in the group were given parts. I was given a part as a policeman. It was only a minor role with a couple of lines of dialogue, and with me then having to chase after someone. But nevertheless I was excited and looking forward to it. I had been given the script but hadn't been given the exact date when my scene would be filmed, but told some time in November, and would later be contacted with the date and details. I received a letter early November with the date that they were going to film my scene. It was to be filmed during the last week of November, and on the same day that we were putting on the first production of plays. I thought, of all the days they could have picked! But as the production was in the evening I tried to see if I could work around it. But in the end had to choose one or the other. I chose the play. I'd put almost three months of work into rehearsals and learned hundreds of lines, and it was more important to me. And also, I would have been letting down other people. I was disappointed with missing out and it would have been a good addition to a C.V. But such is life.

On the night of the production of plays, the evening went well. And the whole thing was captured on video. Performed in the College Hall, we put on four plays and a good size audience turned up. The play I performed in was called 'The Lover': written by

Harold Pinter – a shortened version of it. Playing a character that was performing two roles. The characters I played were excellent parts. And although I had made slow progress in the two year drama course, which the teaching methods were based largely around improvisations, which I had sometimes found difficult, or was uncomfortable with, I always knew in the back of my mind, if given a script, I could take on many roles, and be excellent at them. And by doing this play I had more than proved that; and felt a sense of achievement. It was progress: a milestone.

Another year was coming to an end. I spent Christmas in Wales with my family. And as 1994 dawned I felt a sense of optimism that I hadn't felt for a long time – things had gradually got better. I continued with the production group, and in May we put on another production, a single play which everyone in the group had a part in and as a result of one person pulling out, I had stepped in and played that role too. And by now I had also been nominated to a housing association, interviewed, and put on a waiting list for a flat. I had also been getting regular work working for different painting firms, a lot of it price work, and was making good money. And in June I had moved to a smaller hostel in Vauxhall south London, containing only nine rooms. Living for a year in the main hostel had entitled me to move to a smaller one. And I had concluded during the summer that I had now learned enough about acting to now pursue a career. I had now continually been involved in drama for over two years, and now felt I had gained enough all round experience, skills and confidence, and was confident in my ability, enough to start going for auditions pursuing roles etc. And was in the process of putting together a C.V. and began looking for an agent. And I'd regularly been buying new clothes and had been going out regularly to bars and nightclubs, usually in the West End, and had met and dated a number of women. I was enjoying life again, and things were looking good.

Chapter Four

Voyage

Through Hell

Like I said, things were looking good. But it was towards the end of this same summer of 1994 that I started having problems with my eyes. I had noticed when at work painting, that I was getting pains, in and around my eyes, and at the time thought maybe it was due to the amount of work I was doing, and that it was some kind of eye strain; as it would usually go away when I had finished work. But as the weeks went by it stopped going away, and would often happen when I was reading or looking in the mirror. It seemed to occur whenever I was focusing directly on, or at something. I didn't know what to make of it. I'd mentioned it to some people and my father; who said it sounded like eye strain. I thought in time it would go away, but it didn't; and started to affect my ability to work, and eventually I could no longer work because of it. And by this time I had gone to the doctors, who couldn't diagnose what was causing it either; but had told me about an eye hospital called Moorefield's, which I later went to. And after having my eyes examined by a consultant, was told what the problem was. The diagnosis was called 'Ciliary Muscle Eye Spasm', and was told it occurred probably due to some straining of the eyes. I was given some eye-drops to use three times a day, for three weeks, and was informed they worked by paralysing the eye muscles, and while doing so, allowed them to relax and heal and go back to normal. The side effect from these drops, was that they made the eyes very sensitive to sunlight, and I was also given a pair of very dark sunglasses to wear when outside.

I left the eye hospital feeling relieved. Relieved that they had found out what was wrong, and that there was treatment for it, and thinking to myself, thank God for that. And after completing the three week course of eye drops the eyes did feel a lot better, but were still not a 100% right, and hoping that they would further improve, and in no hurry to take the eye drops again, I first waited to see what would happen over the coming weeks.

This was enough to have to deal with but wasn't my only health problem; a year earlier while doing the cleaning job I had developed problems with my knees, which seemed to affect the tissue under the kneecaps, inflaming it, making them very sore and painful, and would get worse if I walked any distance. It seemed at the time that it was due to the type of work I was doing, as I'd often spent long periods of time working on my knees. I'd seen a doctor and was given anti-inflammatory tablets and later had some physiotherapy on them, and apart from the odd pain here and there, once again they seemed fine. However, during the last two months this problem had started to emerge again and seemed worse than it was before, and eventually I had to go back to the doctor's.

I was referred to the orthopaedic hospital at Portland Square where I'd received physiotherapy a year earlier, and later received an appointment for the first week of November. On the day of the appointment my legs were x-rayed and vigorously examined and the orthopaedic consultant later told me the problem was called 'Chondromalacia Patella' and gave me some insight into what was happening. Chondromalacia is the medical name for degeneration and softening of the cartilage of a joint. 'Patella' is the medical term for kneecap, thus, 'Chondromalacia Patella'. In my case the cartilage that keeps the kneecaps in place had softened or degenerated through the aggravation of being enflamed, causing them to slide out of their proper position. This had affected the alignment of my legs, and the result of this was that some of my leg muscles weren't getting the right exercise; the lack of coordination, had distorted the workings of the legs and joints, shifting more pressure onto some muscles and leaving other muscles unexercised. So when I was walking some muscles weren't getting used and were just getting weaker. It was the muscles on the inside of my legs from the knees upwards that had been affected the most and had gradually become weaker. And the weaker they got the worse the co-ordination became, making the situation worse and even more painful; and limiting my abilities.

Once again I was referred to the physiotherapy department. The aim of which was to learn exercises that would strengthen these weakened muscles and bring them back to their original strength, which would in turn eventually restore balance and the correct co-ordination; and hopefully correct the problem. Starting the physiotherapy a week later I was given a programme of exercises I had to do several times a day, specifically for these muscles. This physiotherapy was different to what had been done a year before, as that consisted mainly of the physiotherapist manipulating the tissue around the knees, to get the kneecaps moving and working properly after the inflamed or damaged tissue had healed badly in and around them -- having a tightening effect.

Meanwhile, my eyes had gradually worsened. And I had no choice but to go back to the eye hospital. I was given the same eye drops to take, this time for two weeks. But after a week of taking them had to stop as they seemed to be making my eyes worse. Sometimes after applying the eye drops I would get severe pains in them. I waited for another week then went back to the eye hospital again. And after explaining the current situation and again having my eyes examined, this examiner concluded that the cause of my eye problems was simply due to the fact that I needed to wear glasses. And that they were straining, simply, because my sight had deteriorated. He put me down for an eye test at the Hospital, but said it would be quicker if I went to see an optician myself as the waiting list was about six weeks. I took his advice and the following day made an appointment with an optician; although was somewhat baffled by his diagnosis that I needed glasses, as I could see fine. But the resulting eye test two days later confirmed this to be true, the optician telling me I needed two pairs of glasses, one for reading and one for distance, and further explaining that wearing them would allow the eyes to see properly and therefore put a stop to the straining.

The idea of having to wear glasses didn't appeal to me at all. But if it was going to cure the problem then I'd have to wear them! I chose the most flattering frames and a week later the last week of November, when they were ready I collected the glasses and put them on. And thought this would be the end of it.

In the meantime I had been enthusiastically doing the leg exercises. It had been three weeks since I started, and progress was slow; and I could only do so much at a time without them becoming painful. I was seeing the physiotherapist once a week, and he was pushing me to do more. And as the weeks went by I was given additional exercises.

By this time I had now been wearing the glasses for a few weeks, but despite this they had made no difference, the problem had not gone away. It was frustrating to say the least.

Some months later: and a new year; and nothing had improved, legs or eyes. In fact my legs had taken a turn for the worse; mainly due to the physiotherapist pushing me too hard by giving me exercises that they weren't ready for: they were too strenuous, but he insisted that I needed to do them. And thinking he knew best I carried on. One exercise in particular consisted of having to lean with my back against a wall, and with my feet about a foot from the bottom of it had to slide my back down, almost into a sitting position, and hold to a count of ten -- several times. And it was while doing this one I had sometimes noticed a slight crunching or grinding noise in my knees; and had mentioned it to the physiotherapist; but was told to keep doing them.

One day while doing this exercise there was a loud grinding noise, followed by severe pain, then, something in my knees just gave way, collapsed; that was it, from that day on they just deteriorated further, and went downhill. It seemed like something had been unhinged, or badly damaged. And the weeks following this they became even worse, and I was now in a situation where I couldn't walk very far without having to take breaks; needing to sit down for a while because of the pain, and could no longer do any of the exercises; my legs couldn't take it.

During this period the physio had given me some tape to strap over my knees, which was supposed to keep the kneecaps firm, in place, and help me to make some progress. I had to tape them up every day and keep this on all day. But it didn't seem to really help, as it often came off or became very lose. And as time went on my legs had become so bad and so weakened that when I was walking they would sometimes buckle, and I now had to use a walking stick.

It was now April 1995, and the physio, not knowing what to do next had referred me back to the orthopaedic consultant. It was a month later that I saw him; and my legs were now in a terrible state, with me only being able to walk some fifty feet at a time, sometimes less, before having to sit down for a break; and was now mainly reliant on taxis to get around. It had created a lot of problems. Just getting out of a chair had become difficult. And on the day of seeing the orthopaedic consultant again, and after once again being x rayed, and sitting in his office after he'd examined me, and again, looking at the results, on him pondering over the results, I'd remarked how angry I was at the state my legs were in. And which I expressed was due mainly to the incompetence of the physiotherapist pushing me so hard with the overly vigorous exercises. And said I felt like getting a cricket wicket and whacking the physio across his knees with it for the way he had messed up my legs. He could see I was angry and didn't say anything to that, but went on to say that he had concluded from the x-rays and examinations and later an MRI Scan that there was no further damage, and although they were in a bad way the only way to get them back on track was with physiotherapy, and said he would refer me back to the physiotherapy department and that it would be a different physiotherapist who would deal with me.

I started weekly visits with the new physio two weeks later, and after an assessment was shown and put through a new programme of exercises, all of which were a lot less strenuous than the previous ones. But as we went through them, some of them caused a lot of pain, and I didn't feel too optimistic about these either, they still seemed more than my legs could take. And as the weeks passed, my efforts to make some progress with them amounted to nothing; due to the amount of pain caused by doing them. Something wasn't right, my legs were in a mess, and I just couldn't see them getting any better under this new programme; my knees just couldn't take the strain of it!

This pattern lasted a couple of months until the physio after having exhausted all her efforts decided to hand me over to the senior physiotherapist who was much older and obviously more experienced. She examined and assessed my situation, and lack of progress, and put it down to the fact that the kneecaps were not being taped up properly; and the tape I was using was no good. It wasn't the right stuff. It wasn't the right tape! The first thing she did was get the right tape, then after shaving the area around the knees set about applying it; and at the same time, showing me, how to apply it properly. This was very important!

There were two sets of tape, and the legs had to be straight when it was applied. The first tape, called 'Mefix' was five inches wide. And a nine inch long strip of this was cut from a roll and put firmly over the knee, horizontally: covering about 3 inches down each side of it too. This acted as a second skin. The second tape was 1 inch wide non stretch plaster tape, from which four nine inch long strips were cut. And two of these were placed on top of this, horizontally, covering and overlapping the top and bottom edge of the kneecap. And in such a way that slightly pulled the kneecap up and held it firm, and in place. Smaller strips were then placed vertically over the ends of this tape as an anchor. And on standing up after this was done, my legs and knees immediately felt much sturdier. I was then shown exercises to do, but not weight-bearing ones, but ones that were done while sitting down. These were much lighter and more appropriate for the state my legs were now in. And as I walked out of the physiotherapy department later that day, I noticed an immediate improvement in how my legs and knees now felt; and could also walk a bit further. This was because the kneecaps were no longer sliding out of their position. They were taped up the way it was supposed to be done, and with the right tape, this had stopped them doing that. It should have been done like this nine months earlier, then my legs would never have got into this state. I blamed this on the incompetence of the physiotherapists I had seen. Now realising also, that you know how bad someone is at their job when you get someone who's good at it. And although there was a long way to go, this at least had seemed like a small turning point.

The same couldn't be said about my eye problems! Now a year since the problem first started, I'd been going through a real hell with them.

It was now the summer of '95. Some good news was that I had got my flat from the Housing Association, and had moved in, in June. It was a one bedroom top floor flat in a street property located in Tufnell Park, North London. It felt good to have my own place. It was just a pity that I was in such a terrible state of health and couldn't make the most of it. I had been given a lot of help moving in and getting furniture and that kind of stuff.

I'd waited two and a half years for the flat. And it was supposed to be something to celebrate. But my health problems ensured that I felt the opposite; as well as restricted and isolated, my eyes had really deteriorated by now. And I had been back and forth to the eye hospital for various appointments for different tests, and had also gone to an eye hospital in Wales, while visiting my family. They had deteriorated to the point where I could no longer read or watch the television or even look in the mirror. Trying to focus my eyes to do any of these was met with severe pains in my eyes, which would sometimes spread from the eyes, around to the sides of my head, and was often followed by throbbing pains. These pains varied. At times some were worse than others depending on what I had tried to look at and for how long. And would only lessen or go away quicker if I closed my eyes for a while. It seemed like the focusing mechanism in my eyes had gone haywire. I could only focus on something for one or two seconds now, and that was it. After that the pain would start. They had also become 'photophobic' (very sensitive to light and bright lights). And this combined with the inability to focus properly had made it now impossible to look at the television; it was just too bright for my eyes, 'overwhelming'. My glasses had been tinted to help with the photophobia, but as far as trying to watch the television was concerned, it was still unbearable.

A lot of things were suggested as to what the cause was. But all the tests that were done on me had found nothing else wrong that could of caused this. I had been given another type of eye drop to try, which were effectively 'false tears' to lubricate the eyes and ease the symptoms. And had also been prescribed stronger glasses. One suggestion was that the first eye drops I'd been given which by now I had used on three occasions had led to the current state of my eyes, i.e. the paralysing effect of the drops had caused the muscles to weaken. And it was suggested I try doing eye exercises, using a pen, to strengthen them.

The last time I actually used those drops was a month after moving into my flat; and which had horrifying consequences'. It was after having had a particular bad phase with my eyes. And in desperation I'd gone back to the Eye Hospital and asked to try them again, hoping that they would take away some of the pains. I was given a week's supply, and after arriving back at my flat applied the eye drops to both eyes, and which five minutes later was met with excruciating pains in both eyes, followed by severe head pains and a dramatic loss of vision; my vision became frosty and hazy, the eye pains worsening, and my eyeballs becoming very tight and feeling like they were frozen. I eventually went and lay down and closed my eyes, hoping it would relieve the symptoms. They did eventually ease, but on opening my eyes again they quickly came back. And my vision had now got even worse, I could hardly see at all; it was frightening! I knew the affects of the drops lasted for six hours, and would have to keep my eyes closed until they wore off, hoping they would then improve. But after the six hours had lapsed nothing had changed, and eventually I ended up having to keep my eyes closed for three days. Staying in my flat with my eyes closed the whole time.

During the three days I had tried to open my eyes a few times but was always overwhelmed with pain. I had some food in the flat and things to drink, and managed to feed myself and make cups of tea and coffee and manage in general, all with my eyes closed. The time I was awake I spent listening to the radio. And it was only on the third day, that I was able to open my eyes again and keep them open, and that some normality had been restored, and my sight gradually came back. It had been a nightmare ordeal and I vowed never to touch them eye drops again.

The eye exercises that were suggested I try, consisted of holding a pen at arm's length and slowly bringing it close to the face, almost to the tip of the nose, while all the time keeping the eyes focused on it: doing this back and forth. I tried doing these exercises, but found doing them to be almost impossible, and found they also made the problem worse. Eye exercises were not the solution.

I'd given a lot of thought as to what else could have caused the eye problems and had eventually concluded that I may have initially started having problems with my eyes due to some medication I had taken for a number of months during the previous summer. It was not long after I'd stopped taking this medication that the eye problems began.

The medication itself was for acne; and was comprised of a synthetic form of Vitamin A. It was called Roacutane, and had to be prescribed by a hospital and only after having tests to make sure you were suitable. And then there was some further monitoring.

I had suffered on and off for years with acne. Not on my face, but across the back of my shoulders and upper back. And from time to time I would get flare-ups. And I had taken a number of medications over the years to deal with it, and was really sick to death of this ongoing scenario, as sometimes it caused a lot of discomfort and sometimes it was very painful. And it didn't look too pleasant either.

I was told about this medication by a doctor who said it had a success rate of 90%, meaning that in 90% of people who had taken it the acne never came back again. Hearing this I asked to be referred to the hospital to see if I was suitable, and after having the necessary tests was deemed so, and told I could start taking it --- which I did for four months.

I remember at the time after picking up the first pack of tablets and reading the information sheet one of the side effects said: 'could cause eye problems, but rare'. I thought nothing more of this and at the time suffered no ill effects during these four months. The only thing I did notice was that my eyes and skin became very dry. But it was right at the end of the course of tablets that I did first start having problems with my eyes, and it seemed logical that this medication had created them, although the eye hospital told me other reasons were the cause. But as the months went by and the different prescribed treatments failed to solve the problem, I had increasingly become convinced that this medication was to blame, and by now had reached the conclusion that it probably was the initial cause: and that the eye drops had made them worse. The Roacutane had certainly had the desired effect on the acne and it never came back. But as far as any regret about taking it; that was too late. Now it was a case of having to deal with these consequences and find a cure, or something that may be able to put right the damage to my eyes I now thought it was responsible for.

But as the year moved on, and in September I turned 30, other health problems began to manifest; namely a stomach problem and pains in my kidneys. The stomach problem I'd had for a while, about nine months, and was told it was excess acid and given medication to control it; first in tablet form, then in liquid. It seemed to vary in symptoms and would be made worse by certain types of food or drinks, and symptoms would range from excessive burping to sometimes being sick, to uncomfortable stomach pains and burning pains.

Over this last number of months it had gradually got worse. This I was told was because of stress created by the other health problems, and the fact that I wasn't very mobile due to my legs, this had affected my digestion I was told! The medication didn't seem to help much with the symptoms anymore, and sometimes after taking it made them worse. It had now reached a point where this was causing me great discomfort and seemed like further investigation was needed, as I wasn't fully convinced it was stress alone or a lack of mobility that had made it as bad as it had now become; and there were new pains emerging that I hadn't noticed before. I'd already had tests to see if I had ulcers, and they came back 'negative'. And on my last visit to the doctor's in early October, I was told to take a larger dose of the liquid medication more often, which I did.

Altogether I had a lot to cope with. And despite mastering the technique of taping my knees properly, I'd made virtually no progress with my legs. And again it was due to pain brought on by trying to make progress. Even these light exercises were causing problems, I couldn't win and was gradually becoming more and more depressed. A good thing about being in the hostel was that there was a communal area and people to talk to. But now that I had my flat, I was on my own. But this on its own didn't bother me. It was the isolation, brought about by being incapacitated, which was frustrating and depressing. This and not being able to read anything or even watch the TV, and not having the proper function of my eyes, which, combined with being alone had led to this growing feeling of isolation; not to mention the physical pain I was having to endure.

Over time, the stomach problem had become more severe; and had reached boiling point one evening during the last week of October. I was in that much pain I felt that I had no alternative but to go to hospital. I called a taxi and had it take me to the Accident & Emergency Department at the Whittington Hospital in Highgate. And, after the procedure of giving the necessary details, waited to be seen. And after two hours my name was called and I was taken to a room to be examined. And after some tests and a physical examination were done I waited for the results.

A lot of time passed before a nurse came back with the results; and then told me that they had found nothing else wrong with my stomach, and that I was free to go.

This was hard to believe! I couldn't believe that I was in so much pain and that they couldn't find anything wrong. And the thought of having to go back to my flat in this state was unbearable. The last couple of weeks I had become so severely depressed about my situation, I'd felt suicidal. This stomach problem was pushing me over the edge.

I told this to the nurse, expressing how bad things were and said there's no way I can go back to my flat tonight I feel like killing myself that's how bad I feel. She then asked if I would like to see the psychiatrist. I was slightly puzzled, as to why she'd asked if I wanted to see a psychiatrist, as it was my stomach that had been the problem and had brought me to the hospital, but said yeah anyway, as I didn't want to leave. Maybe I could get a bed for the night or something; anything but go home. She said I'd have to wait a while -- which was fine by me.

It was now midnight. I waited in the room, sat in a wheelchair that had been used to bring me from the waiting area, as I was still in the situation of being able to walk only some 50 foot at a time before needing a rest, and had requested the wheelchair after noticing their area of operations was at the end of a long corridor, quite a distance from the waiting area – further than I could walk. The psychiatrist arrived around 2.a.m and introduced himself, then began by asking me to tell him about my situation and what the problem was. I told him how I was feeling, and elaborated on the reasons why. He then asked about my past, and asked me to talk a bit about that. Then there were some more questions. And after about 20 minutes he said, I would like you to come into hospital, which I was glad to hear him say; and agreed that it would be helpful. He said he would arrange for me to be taken to a ward and then left. And shortly thereafter I was taken by a nurse and security guard to a building separate from the main hospital and up onto a ward.

Although it was now probably 3 in the morning, there were people about, obviously patients; and I soon realized where I was. When the psychiatrist said he wanted me to come into hospital, at the time I didn't realize he meant the psychiatric unit: but this is where I was. And although I went through the process of being signed in, I didn't feel too enthusiastic about staying. I had somehow been under the impression that I was coming into hospital for further observations on my stomach just in case there was something they had missed, not wanting to take any chances, this kind of thing; and thinking that the psychiatrist was just making sure I did have a problem with my stomach, and it wasn't psychological. The last place I expected to be put was in a psychiatric unit.

After being signed in I was given a sleeping pill and a bed and told I would be seen by someone in the morning. I was feeling very tired by now and was just glad to get into a bed and by now didn't particularly care where it was. The stomach pain had also eased, the painful symptoms considerably diminished. I soon fell asleep.

Waking up later that morning, I was introduced to a key worker, for any help I may need, and given a locker and informed about the routine. And was informed I wasn't allowed to leave the ward for 48 hours, until some kind of assessment had been done.

The ward itself was one half male the other half female. And each half split into a number of smaller wards, each of which contained six to eight beds. And for the whole ward there was a communal room with a television, where you could make yourself a cup of tea or coffee. And this seemed to be where many of the patients would congregate, and where I soon found myself joining them. And although I'd had initial reservations about staying, as the next few days unfolded my attitude changed. And although there were some very disturbed patients, there were also those you could have a conversation with, and whose reasons for being there ranged from manic depression to schizophrenia, to people who had attempted suicide.

I had not realized that it was a psychiatric unit where they treated people for depression and those who felt suicidal, and had assumed a psychiatric unit was for the severely unstable, as did many people. This was the general perception! But after talking to some of these patients, it dawned on me why I had been brought here. And one of the reasons was that I'd expressed that I was in danger of killing myself. I could now see why the nurse had asked me if I wanted to see a psychiatrist.

My other reasons for a change in attitude was that I felt I had no choice but to stay, I needed to be here, I was in such a bad way myself, and if anything bad happened to my stomach at least I was in a hospital. And it was a nice change to be around some people, have some company.

It was on the fourth day of being here that I was taken in to see the psychiatrist. Still using the wheelchair, I was wheeled into a room and confronted by a dozen or so people sat in a half circle in what was termed the ward round, i.e. the weekly gathering of the Mental Health Team for that ward headed by the psychiatrist, to assess the progress of patients, and diagnose new ones for suitable treatment. The psychiatrist - and he was not the one I had seen initially - asked me what was wrong and why I was in a wheelchair. I told him how depressed I'd become due to my health problems, telling him about my eyes, stomach, and then my legs, the name of the problem and how it caused the muscle strength to degenerate, and explained the problems I was having trying to build them back up again. He was quite abrasive in his manner, saying, forget all this stuff about muscles; you need to get out of that wheelchair and start walking. Taken aback by what he said, I didn't know what to say! Thinking only, that he had no idea what I'd been going through with my legs, and it was easier said than done. Plus the wheelchair was helping me get around. He then went on, saying he would give me anti-depressants for the depression; and that was the end of it.

Later that day a nurse said that they wanted to take the wheelchair off me, saying it was for my own good. I was reluctant to part with it, and refused, stressing how much I really needed it; but that night while asleep they took it.

Two days later I was moved to another ward on the next floor up. There were three wards altogether and you were put on the ward that covered your area postcode. This was the one that covered mine. The other ward had been temporary until a bed had become available. This was a much nicer ward: painted in a different colour, and had a much calmer feel to it. Everything about it was nicer, including the staff. But although the psychiatrist had prescribed anti-depressants, I'd give it some thought, and decided mainly because of side effects that I didn't want to take any more tablets, and had told them this. And after a few days I also saw the psychiatrist for this ward and told him the same thing. A couple of days later I had an hour long interview with the ward doctor, as well as asked about my physical health problems, I was given a physical examination, and asked a lot of questions to do with assessing my mental state. And although I told her how severely depressed I'd become, and why, I was careful about what I said to some of the other questions to do with mental health. And this was because of patients I had talked to, who had been what was termed, 'Sectioned under the Mental Health Act,' i.e. not allowed to leave for an indefinite set period of time. You were kept there as long as necessary if it was deemed that you were either a danger to the general public, or yourself. Just one example was a man in my ward who'd been kept there for nine months. He was just one of many who had been sectioned under this act.

The overall impression I got was you were more likely to be sectioned if you posed a threat to the general public. And once under a section, medication was forced on you.

And the reason I was careful about what I said was that for quite a number of years, I had had things and thoughts of an unusual nature coming into my head which I couldn't explain, and which had gradually got worse in their content. And although I knew the thoughts were not my own thoughts I couldn't work out where they came from either. But although they were unpleasant and often disturbing, decided it's probably better if I don't mention them, as I could end up being sectioned myself.

I had first noticed these thoughts in my early 20s when 20, 21 years of age, and when I got to the age of 23 remember specific situations where images and thoughts appeared or flashed into my mind, and were of such an evil or bizarre nature, remember thinking, where's that coming from, that's not me, and being quite unsettled by them. But by now I had become used to them and just kept it to myself, not really wanting to tell anyone about them, as people would probably think they were my own thoughts and avoid me; or worse, I would be locked up. So I thought for the time being it's better if I keep it quiet.

Anyway, by the time the interview had ended, the doctor had managed to talk me round to the benefits of taking the anti-depressants, and convince me that they had very little side effects, and taking them would help me a great deal. And convinced by what she had said I decided I would give them a try.

I started the medication the following day. It was called Paroxetene and was prescribed for anxiety and depression; and was informed the full benefits would not be felt straight away. The immediate benefits I felt were that it made me feel relaxed and it took my mind of things. I just felt a bit more relaxed about everything. And the stomach problem although it seemed to fluctuate in its severity, between severe and not so bad, on the whole it had become more manageable. And to see how things were going, a week later I was seen again by the ward doctor, who remarked that I was much calmer. And although I had still been trying to improve my legs with the exercises, it was around this time that I did start making progress. And this was due in part to a long table in a hallway on the ward. I had started using it to do my leg exercises on, as it was more convenient. Otherwise I had to somehow get onto the floor to do them; as I had to be in a sitting position with the legs outstretched straight, and on a flat surface.

At this point the muscles affected had wasted away that much that there was only two exercises I could do. One consisted of turning the foot inwards and then tensing the inner thigh muscle from the knee upwards and holding it to a count of ten. The other was: putting a

44

couple of books under the joint below the knee and then raising the foot off the surface, straightening the leg; doing this a few times up and down. I could manage a few of each, and because of the convenience of the table I was doing them almost every hour. And as the days rolled by I was able to do a little bit more. These exercises were bearable and without much pain. I think to a large degree the fact that the knees were now being taped up in the right way, and the kneecaps were not sliding out of place, in time, had allowed the inflammation and any damage to settle down or heal, and that's why I was getting a lot less pain. Only a couple of months earlier I was trying to these same exercises, and in order to do them was having to put ice packs over and around my knees, as they would often flare up, causing a great deal of pain. But now the ice packs were no longer necessary, and I could do more of them.

I had now been here three weeks; and was starting to feel a bit better. I had been thinking about how my fortunes had turned for the worse. Just over a year earlier I'd been planning to pursue a career in the film business. And now I was incapacitated, and in a Psychiatric Unit. At this point those aspirations seemed another world away; and concluded, for the time being it was best to just forget about it; as what lay ahead wasn't predictable.

As the weeks went by things continued to improve, my mood was now much better, and I was no longer experiencing feelings of severe depression. And overall was feeling more optimistic about things. The medication had obviously played a large part in this. I was still in a physical mess, which there was no escape from. But had to persevere and build on any improvements I'd made with my physical problems. That's all I could do really.

I was discharged the first week of January '96, after just over two months. I'd been on a few home visits to my flat prior to this to check it was ok, and to get used to being back there again. But that wasn't the end of it. They had referred me to the Psychiatric Day Hospital for further treatment and help. This was on the first floor in the same building, and I was informed the wait would be a couple of months. In the meantime while I waited I'd had a scan on my stomach. And although my stomach in general had been better, it was still causing me a lot of misery. This scan was called a 'Barium

Meal.' A more sophisticated x-ray type scan which would detect or show up anything that was wrong with it, or causing the problem: the results were sent to my doctor; and came back negative – no abnormalities had been found. He concluded that at some stage I'd probably had an ulcer, and it had healed. Or the excess acid may have burnt a hole in the stomach lining and had also healed. For the existing problem which he said was excess acid and was causing the problems, he gave me a new medication to take. It was called Losec, and proved to be the one that eventually brought the problem under control. And after only four months of taking it the stomach problem had virtually gone away. It was a good drug, with what seemed like no side effects.

And, while I had been on the ward I had also noticed a good side effect of the anti-depressants. The relaxing tranquillizing effect had somehow benefitted my eyes. It had probably relaxed the eye muscles, and in turn obviously helped them function better. Whatever it had done it had improved them. I first noticed this a couple of weeks after I started taking it. I had caught a glimpse of the television screen on the ward and was surprised that I could look at it without getting the usual painful symptoms: the light sensitivity seemed to have been greatly reduced. And over the following weeks there were further improvements on the other eye symptoms, the focusing etc. And now that I was back in my flat, instead of just listening to the television I was able to watch it again; although in a limited capacity, i.e. I could only look at the screen for a few seconds, just fleeting glimpses, as there was still no real power in my focusing. But even in this capacity it was enjoyable to be able to watch the television again after not being able to for over a year. This alone helped me feel less isolated.

A couple of months had now gone by since being discharged from the Psychiatric Unit, and overall everything had improved. And in March I'd had an assessment type interview for a place at the Day Hospital and then started there two weeks later. Attendance was from 9-30 till 4-30 pm, Monday to Friday.

The daily routine at the Day Hospital was structured around a programme that had been put together specifically for each patient, depending on their illness or what they were suffering from, and what would be beneficial to improve, help, or make them well again. It comprised of some group therapy classes of a psychological nature, and also some occupational therapy --- Pottery, Woodwork, Art, Cookery, Yoga and Relaxation etc. And a patient's programme would consist of some of each: both occupational and group therapy. You were also assigned a care coordinator who was responsible for your care while there, and reporting on your progress, and who you had to see for one hour once every week; and every so often you were seen by the Psychiatrist, in what was a very similar setup to the ward round on the wards. There was also a once a week community meeting of Staff and Patients to discuss any problems, announce up and coming classes or events, and to introduce new patients and staff.

My initial start comprised of a written report being done on me by my care coordinator, Barbara Richardson, who was in her mid thirties and studying to be a Psychiatrist; and was at the Day Hospital for six months as part of her training. The report was done in three sessions over a three week period, and seemed to cover my whole background up to the present. My initial start there was for two days a week, then after three weeks, every day, Monday to Friday.

I suppose as well as a place for treating people with mental health problems, it was also a safe environment to be treated in – with support. The type of patients at the Day Hospital, were suffering from something obviously; although not to a degree that they would need to be an inpatient on one of the wards. They could receive treatment as a day patient and this would be a more suitable environment, for a number of reasons. There were about 30 or 40 day Patients. And some like me who had been an inpatient and were later referred to the Day Hospital for further treatment, or monitoring: after their illness or situation had stabilized, to a degree that they no longer needed to be on the ward. But generally most Patients had been referred by their doctor.

My first few months were spent mostly doing occupational therapy: 'Pottery, Woodwork and Relaxation'. And despite the problems with my eyes and legs, with help, managed to participate. And any free time was spent sitting in the common room, initially, mostly on my own, away from the other day patients; in a corner, where I had felt more comfortable, and could also do my leg exercises.

The common room, was a large open room type hall, and which a section of was set aside for the day patients to convalesce, during times they didn't have groups or anything to do, usually dinner time between 12-30 and 2 o'clock and at certain times morning and afternoon, when tea and biscuits was served in the canteen. It also had a door that opened out to a small garden area with benches and a pond. The canteen which was next door was where day patients as well as inpatients had their meals, the inpatients having theirs first, lining up in the common room at 12 o'clock; and afterwards some of them would linger around in the common room, or sit out in the garden area, before having to go back on to the ward -- a routine I was familiar with. The day patients would go in at 1 o'clock.

Over the following months, I had gradually felt more confident, and able to talk about some of the disturbing things that came into my head, as well as how sometimes they had affected my behaviour. Now feeling it was safer to talk about this: Mentioning it initially to my care-coordinator. I'd felt for some time that something wasn't quite right and it bothered me. Now feeling that maybe there was a malfunction in my brain or something.

And later in August of this year I was sent for an appointment at the Royal Free Hospital in Hampstead. To see a Psychiatrist, who specialized in a specific area of Psychiatry. I answered all of his questions. I didn't avoid anything, and elaborated where I needed to. And later when the results of his diagnosis were sent to the Day Hospital Psychiatrist, I was put on an antipsychotic drug, called Sulpiride. Only a couple of months earlier, I'd been sent for a brain scan. I think they were probably just checking to see if there were any abnormalities, in order to rule that out, which their weren't, it came back normal.

It was now late September of 96, and Barbara my care coordinator had recently completed her six months and moved on. I was sad to see her leave, as I was fond of her and had got use to the weekly hourly meetings. Her other patients felt the same. And she was well liked and well thought of by all the patients; probably because she had a genuine lovely warm manner about her. I remember saying to some of the patients after she'd left that she reminded me of Princess Diana, and they all agreed straight away that the comparison was spot on. I had now been assigned a new care coordinator named Mike, who was also a trained Psychiatric Nurse, and had a permanent position in the Day Hospital. It wasn't easy going from one person to having to deal with a new one – it took a while to feel comfortable. But he turned out to be very good.

The year was now coming to an end, and slowly things seemed like they were improving, and the following year started off a bit brighter.

It was now 1997. And I was still a patient at the Day Hospital. I had been doing well with my physical health, making steady progress with the leg exercises, and had progressed to doing more vigorous ones. It was a slow process and I had to be careful not to overdo it, as I had done this on a few occasions and it had badly enflamed everything and had set me back. But I could walk a lot further now, and it seemed like I was now on the right path. The eyes: this was more frustrating, and also depressing. And although there had been some improvement I was still limited in their use in what I could do. Two months after I had initially started at the Day Hospital, an appointment had also been arranged for me with the Eye Department in the main Hospital; where I had been through a series of tests and examinations. They found no abnormalities with the main components of my eyes, only a lack of convergence (ability to focus properly) and had also suggested eye exercises for it using a pen, to get the muscles working better; which I had already tried and found them to be almost impossible to do. I gave them another try anyway, and after a month of trying gave up on them; and for the same reasons as before.

I'd grown tired of seeing all the different people in the pursuit of trying to get my eyes better, I had needed a break from it all and had decided to just leave them alone for a while after this, and see what happens, maybe in time they would improve on their own. Eight months had now passed, since making that decision, 'and the eyes had been on my mind again. I'd been thinking about the amount of times I'd had my eyes looked at; and although there had been some more improvement and I was able to focus on things for longer periods, and could now read a little bit as well; they were still a long way from cured. And although I had drawn my own conclusions as to why my eyes were in their present state, I couldn't help but think, that surely, there's something they must have missed, and something's obviously wrong somewhere. And, decided to find out what other areas of expertise were available. I thought if going private paying to see a top eye specialist would be better; thinking that maybe they might know a bit more, and have a broader range of experience, and would able to pinpoint or diagnose what had gone wrong, and come up with a solution to correct it.

I decided to pay my doctor a visit to ask about any top Eye Specialists that he could put me in touch with. He gave me the names of two, one of which, and nearest, was based in the famous Harley Street. I choose the one in Harley Street, and promptly phoned for an appointment and got one for the first week of April.

Meanwhile: I had now been taking the 'anti-psychotic' medication for about six months, and had not noticed any real benefit. Only, that after taking the larger dose at night, that it helped me get to sleep quicker. The negative effect was that it sometimes made me feel a bit drowsy for a couple of hours, after taking the morning dose.

The first week of April soon arrived and I went for my appointment in Harley Street, hopeful of a good outcome. But what would result from this would be horrendous consequences for my eyes. The Eye Specialist found no damage to any part of my eyes and only what he described, as a weakness in the muscles, which was responsible for the lack of convergence. His examination and tests were meticulous, and his advice, was not to wear the glasses all of the time, as he said that they can restrict or cause the eye muscles to weaken. He said taking the glasses off for 15 or 20 minutes several times a day, in my case would get the muscles working better and help to strengthen them, and that would improve the overall problem. His advice seemed logical and I hadn't come across it before and thought it was worth trying, and started immediately the minute I left his office. Walking out into the street the sun was shining, it being a nice spring day, and I took off the glasses. And it felt strange to be walking about without them on, as I'd worn them from when I got up till I went to bed, every day since I'd been prescribed them 18 months earlier. I did what he said, keeping them off for 20 minutes then putting them back on. I did this every couple of hours, then after a week or so, I was taking them off more frequently and for longer periods of time, only putting them back on, if my eyes began to hurt or strain. It seemed to be going well. After just over a week, they already felt better and stronger and my sight had even improved -- my eyes were improving fast. And after two weeks there was even more improvement, seeming like finally my eyes were now on the right track too.

But this unexpected improvement was short-lived: as things took a horrendous turn for the worse. It had been just over two weeks, when one day standing in my kitchen I'd had the glasses off for a while and felt my eyes straining a bit, I put them back on, but after a few minutes something in my eyes changed, followed by very bad pains, and then my vision changed, going fuzzy, and suddenly all objects around me looked much smaller. I thought, what's happened now! The pain increased. And not knowing what to do I took the glasses off. And as I looked around I noticed my sight' had also deteriorated. I went and lay down for a while, and kept my eyes closed; thinking, it was a temporary thing and would soon clear up and go away. The pains gradually went, and after an hour my eyes felt better; and thinking the problem had cleared up I opened my eyes. But nothing had changed, only the initial pains had gone, my sight remained in the same state, and I soon felt my eyes straining too, which was unusual now: because I'd been keeping the glasses off for much longer periods of time, and this particular kind of straining was symptomatic with needing to put the glasses back on, to ease it: 'this straining now occurring after only a few minutes.' The pain becoming so bad, I had no choice but to put the glasses back on, and which eased it, some of this straining gradually going away.

I don't know what had happened, but whatever it was; it was a big setback, and depressing. But I was also soon made aware that the photophobia had returned too; as I found out later when I switched the television on, once again it was unbearable to look at. This was especially depressing. All I could do was see what happens over the coming days, and during which I tried carrying on with the same technique, taking the glasses off for a while etc. But after only a few minutes of doing this, the straining pains quickly emerged and I had to put them back on. It seemed that all the improvements I'd made had disappeared; and my eyes were now back at square one. But as the days wore on, I'd also worked out what had happened. 'The sight had obviously improved', so much so, and very quickly, that the glasses gradually became too strong, stronger than was needed, and with the eye muscles still in a weakened state, had buckled or contorted in some way under the pressure of having stronger than needed lenses being placed over them. On that particular day I'd kept the glasses off for two hours and during that time there may have been additional improvement, that when I put the glasses back on it was by now too much for the eyes, to strong, and it caused a bad reaction. It couldn't have been anything else. Things had improved that quick that it hadn't occurred to me that I would need to get a weaker pair of glasses. And that had been the big mistake. Now my sight had gone back to the way it was before, and with worse symptoms. One being the photophobia, which by the end of the week had become even worse: and seemed ten times worse than it was before. Everything looked abnormally bright, particularly the colour white. It was like looking at a fluorescent white light. And I had to avoid looking directly at certain things because of it. Even a sheet of white paper looked abnormally bright. I also had to cover all the light shades in my flat as the glare of light bulbs was now too much for my eyes. And the television screen had to be covered also, as it produced the same reaction. And over the next couple of weeks life again had become a real hell. It seemed like my eyes were now worse than ever. And being outside was very difficult; especially at night having to deal with the car headlights, which were now unbearable to look at: bright lights would cause the worse pain; which were sharp type pains often lasting 5, 10, 15 minutes. And as well as painful they were demoralizing and stressful; and it was important to avoid looking at lights of any kind as they had this same effect. This recent addition, of a more severe strain of the photophobia to my eye problems had created an obstacle course. And I wondered how long this would last before it would get a bit better.

But, it didn't seem to get any better. This severe sensitivity to lights and other things stayed the same.

I'd remembered and had been aware, that a side effect of the anti-depressants was that they caused the eyes to be more sensitive to light, but while on them they hadn't caused me any problems but had actually helped my eyes. But now I thought maybe if I stopped taking them for a while it may take away some of the sensitivity or the sharpness of it. Maybe they were now contributing to the problem. I was desperate to try anything that would reduce it and decided to stop taking them in the hope that it would help. And although it was a bad situation, I avoided getting darker tints put on my glasses, deciding first to wait for a while to see if an improvement would occur; mainly because I was informed that having them to dark can gradually make the eyes even more sensitive to light, so I waited. But a couple of weeks later nothing had changed; only that I'd become very depressed about the whole situation and the aggravation I was having to deal with. And the frustration of having made so much progress with my eyes, and now they were in a worse state than ever. And it wasn't just the severity of the photophobia! But the focusing mechanism had gone very bad again also. And once again my eyes had turned my life upside down. The eyes are everything. And when something goes wrong with them it affects almost everything you do. My mood had also gradually plummeted, now often spiralling into deeper depressions, consisting of thoughts of suicide, and feeling drained. It got so bad, my mood had spiralled so much so I was seriously thinking of killing myself, and thinking of the best way to do it; I'd had enough of it all, and just couldn't see my eyes getting better again. And having to keep living like this was a nightmare.

It had now been well over a month since this latest scenario had begun; and my care coordinator at the Day Hospital had seen my decline. And recognised

that things had become that bad, that I needed to be back in hospital; on the ward, for my own safety. And had said this during our weekly meeting: and that if I agreed to it he'd arrange a bed for me.

I certainly wasn't expecting to end up back on the ward again, but felt in the circumstances it was probably the best thing for me, and agreed to it, and that afternoon went home to pack a bag of necessity's, and when I came back a bed had been arranged and I was taken up to the ward.

As I had been here before I soon got settled in. Most of my first week was spent languishing on my bed, which was next to a window. And dwelling on and thinking about the hell I was having to endure, often feeling tortured by it. The amount of times I'd regretted taking the acne medication, Roacutane, and wishing I could turn the clock back. But I couldn't, this was it, this was my reality. I had very little will to really do anything; and my appetite by now had almost vanished.

I was soon put back on the anti-depressants, my stopping taking them had had no benefit on my eyes and from what I gathered had only contributed to the severe depressions that I was now experiencing. My care coordinator had told me this during our last meeting, saying that just stopping them was dangerous; it had to be done gradually. It was only during this meeting that I had mentioned that I had stopped taking them, although by this time it had been almost three weeks since I had done so, and I think this also was why he had made the decision to get me back on the ward. On the plus side of being here, it was a badly needed break from the madness of it all, and where my needs were catered for. This time, my stay on the ward lasted four months. It was a similar scenario as the last time I had been here; ups and downs then gradual improvements. But in many ways this time it seemed worse.

I was discharged the middle of August, and went straight back to the Day Hospital. And the only improvements in my eyes' were due to having got my glasses darkened with darker tints; which in the end had felt I had no choice but to do this, as the brightness of some things were so bad. But they still remained in the same state; and going back on the anti-depressants had not benefitted my eyes in the same way they had the first time I took them either - the same improvements did not occur. It was only my acceptance of these new symptoms, and having to adapt to them that had changed, any bright lights were still problematic; they still managed to pierce through the dark tints and cause pains. But were not as bad as before; the darker tints taking away some of the sharpness. I could now no longer look in the mirror again either, the moment I tried to focus on myself in the mirror the reaction of my eyes was they went haywire. And because of the pains it caused in and around my eyes, gradually I gave up trying; preferring to do without it. Like I said, I just adapted, finding other ways of doing things. And I was also back to the situation of needing other people to read my mail for me, as I could no longer do it myself. Even using a magnifying glass was not an option, as the strength of these was too much for the eye muscles. And once again, not being able to watch the television, I now listened to it instead – after taping cardboard over the screen. But gradually as time went on I got more into a habit of listening to the radio, and generally found it to be more interesting. There was a good range of stuff on the radio, from phone in shows, to documentaries and news, to music stations. And over time I tended to listen to the radio more, as I found it more stimulating, and most of the time with my eyes closed as this would give my eyes a rest. And generally began to find it more pleasant and I noticed it also allowed me to think more deeply or in a more concentrated way; almost drawing on something that wasn't there when the eyes were open.

Not long after I had been back at the Day Hospital; during a weekly meeting with my care coordinator, he had mentioned and asked if I'd been told that I was suffering from Schizophrenia. And that this,' had been the diagnosis after my interview with the Psychiatrist at the Royal Free Hospital.

'I knew that I was probably suffering from something of a Psychotic nature', by the very fact I had been given an anti-psychotic drug. But hadn't realized that that was the diagnosis. I certainly didn't think I was mad or mentally ill in that way; or insane. But I knew that there seemed to be something that wasn't quite right as well. And on hearing this, at the time, I thought to myself that maybe I've just got a mild case of this Schizophrenia; and didn't know what to make of it all really. And with everything else going on I just took it in my stride. It didn't seem that important.

It had only been a couple of months since I'd left the ward and came back to the Day Hospital. But it wasn't long before another health problem developed: this time with my hearing. It hadn't been that long since I'd recovered from the last batch of misery and hell, and now, I was having to deal with something else. This had emerged one Saturday night in my flat while I was sat listening to music on my music system and sitting very close to one of the big speakers. It was on, not excessively loud, but moderately loud. And as I had been sat there, I had noticed that all of a sudden the music seemed to get louder; and then got very loud – although I hadn't turned it up. I thought that's a bit strange and felt forced to put my fingers in my ears for a few seconds, thinking it may just be something temporary, a fault in the system or with the CD or something, and would soon correct itself. After about ten seconds I pulled out my fingers from my ears; and which were met with a blast of noise that seemed even louder. And the effects of this on my hearing soon surfaced; in the form of loud ringing and then burning sensations in my ears, which were also painful. I turned the music off and later went to bed, thinking by the morning it would have gone. And on waking the following day these ringing and burning sensations did seem to have disappeared. But later that same day while in a cafe I noticed that the music that was playing, although on low, it seemed too much for my ears, and I could feel the inside of them burning; and also noticed that everything in general seemed louder than normal. And later when I got home, when I turned on the radio the same thing happened, the burning pain started again, then followed by ringing, 'then throbbing.' And after half an hour I had to turn the radio off, even though it was only on low; as the symptoms were getting worse. It seemed in part to be a reaction to the conversion of sound "coming through a speaker," maybe the static in it. I left it a few days, and then after no improvement went to see my doctor. I explained what had happened, and the symptoms; and his diagnosis was that I had bruised my ears, and said it would probably be better in a couple of weeks, and gave me some anti-inflammatory tablets, to help heal the bruising; as he called it.

Three weeks later and it hadn't got any better – but worse. I went back, and he then referred me to the Royal Throat Nose and Ear

Hospital in Kings Cross. Two weeks later I received a letter from them for an appointment with a Consultant. But the appointment was five months away; not until April of 1998. I called to see if I could get an appointment sooner, but there was none available. But was told I would also be put on the cancelations list, which sounded more promising.

In the meantime, the full-scale effects of this problem began to emerge. Something in my hearing had changed, something had been disturbed. And as the weeks went by the results of this were that I could no longer listen to the radio or television, go into anywhere where music was played, or a radio was on, as it seemed that any sound that came through a speaker even on low volume, seemed overwhelming and too much, and had an adverse effect on my ears; and would trigger off these symptoms, enflaming them. Anything on low was bad enough, but if something was loud it was totally unbearable and I had to remove myself from the situation really quick. And police and fire engine sirens were now unbearable to my ears, and when I heard them approaching I had to quickly put my fingers in my ears, as they couldn't withstand this volume and type of sound.

It was an unusual and horrible problem. And I found throughout a day I would have to put my fingers in my ears for a whole number of sounds that were too much for them. And the world now seemed like a very noisy place. I tried to cope with this by using foam earplugs which eventually I found myself wearing all the time. And although they seemed to help in general – for certain noises I still had to put my fingers in my ears, and this was with the earplugs already in them. Even talking on the phone was difficult, I had to keep the phone away from my ear and keep phone conversations very short. So while I waited for this appointment at the Ear Hospital, I also decided to see what was available in the Alternative Health Profession, that may be of help.

At a Healing Arts Exhibition I'd visited a year earlier in Victoria, someone had given me a leaflet about 'spiritual healing'. It was from an organization called the National Federation of Spiritual Healers, and gave a brief account of what spiritual healing was. I had kept the leaflet and had recently thought about calling them. And now after deciding to do so, was given a list of healers in my area, and then called one of them, and started going regularly for spiritual healing.

I'd actually seen a healer a year before, called Phil Edwards, a well known spiritual healer, who I'd heard one night on a radio programme, being interviewed about many of his healings. I'd been to see him three times. At the time hoping he could cure my eyes; but it didn't happen. But nevertheless, I kept an open mind about spiritual healing. And in January of the New Year, now 1998; I also got in touch with a place called the White Eagle Lodge. A spiritual organization someone had told me about, and was based in Kensington, South London, and who also did a form of spiritual healing: using different colours for different ailments. And in February I also started going there for healing, usually once a week.

Spiritual healing is the channelling of a divine source of energy into the recipient. And it supposedly goes to where it's most needed; this is what I was told. And with the different types of spiritual healing, they are more or less a manipulation around this same source of energy.

I had also gathered by now that this spiritual healing didn't always cure things straight away, but took time, it was a process. And I was prepared to stick with it for a while and see what happens. With the different things I had wrong with me it seemed like the reasonable thing to do.

I hadn't received an earlier appointment from being on the cancellations list. And in the end did have to wait the full five months to get my ears looked at. And by the time of the appointment in April the sensitivity of my ears was even worse. And on the day; and after being examined and undergoing a series of tests, the Consultant had also said, that my wearing the earplugs all of the time was making the sensitivity worse: making the ears even more sensitive, 'and told me not to wear them all the time. At this point it now seemed that I couldn't get by without them. But what she said was right as they hadn't seemed quite this severe a few months earlier. And after she had gone through and completed everything said I would now be referred to a Hearing Therapist' who would help me get my ears right again. And this would be another long wait: The appointment I received was for three months later.

While I waited for that, I started the process of getting rid of the earplugs. Starting by going to a nearby park in Highgate, and while there taking them out, as it was very quiet. This was to let them get use to hearing again without the earplugs. And to reduce the extra sensitivity that had accumulated because of their constant use. I would generally stay there for a few hours! Everywhere else was just too loud and too fierce on the ears: the noises of traffic etc. I couldn't just take them out in my flat either as the noise of the traffic going by was similar, that's how bad they had become. So the park was the best place to start.

To give an indication of how loud things had become; one day I was sat in the park and there were two people sat on a bench having a quiet conversation about a hundred foot away; and I could clearly hear everything they were saying. At this point I would have made a good spy.

When the day of the appointment to see the Hearing Therapist arrived, I was then told what the problem was called. And by now had gradually weaned myself off needing the earplugs to the degree, that most of the time I was not using them. And there had been a good reduction in the additional loudness and sensitivity that had developed through there constant use. The Hearing Therapist' said the problem was called 'Hyperacusis', in other words, hypersensitive hearing, and explained what had happened to my ears.

Firstly, I was surprised there was a name for it, and wondered why the Consultant had not told me this, and was very annoyed; mainly because of the anxiety and stress I'd endured because of not knowing what was wrong, or if it was curable. I'd even tried other forms of alternative medicine because of this, namely: Homeopathy, Chinese Medicine and Acupuncture, as well as the spiritual healing. It was a horrible problem, and I had been looking into anything that may cure it. In all, 'I had waited eight months to be told that.

And what had actually happened to my hearing, put simply, was the hearing mechanism had been knocked out of balance. And the method for curing this was with the use of a device called a 'White Noise Generator. Which looked like a hearing aid,' and was attached to the ears in the same way. 'One for each ear, and a tube connected to them was put in the ear. When switched on it made a noise like the sound of air coming through a hose. This noise was the combination of all the different frequencies of sound at a low level; some of which my hearing could no longer withstand.

The idea of the white noise generators was that the gradual increasing use of them, using all these frequencies' together, would retrain the hearing to be able to accept these sounds again. I was given instructions on how to use it. To start with I was to use them for half an hour each day, then over time as my tolerance grew it would be increased. I was told it would take time to cure and would be seeing the Hearing Therapist regularly as she would be coaching me along the way. I was given an information sheet on Hyperacusis, and another appointment for a fortnight's time; and off I went.

It now seemed like a large part of my time, was taken up going to various appointments to do with my health and trying to get better, whether through conventional medicine or by alternative means. And only a month later I'd also had to go to the local hospital accident and emergency department and spent most of the night there, suffering the effects of a kidney stone; which are extremely painful. I had had one of these before. In June of 96 while visiting my family in Wales, I was taken into hospital and stayed there for a week. That was the first one, and the worst. The pain was something else! I've never felt such agonizing pain ever; it was that bad I wanted to die. While there I was given morphine for the pain, but even after that the pain was still bad. I was put on a drip, and later during that week the stone eventually worked its way out of my system. I remember later a nurse saying to me that the pain of a kidney stone was equivalent to the pain of giving birth, and I had said, oh I think a kidney stones worse. So that gives you some idea of the type of pain.

This second kidney stone was agonizing but not as bad as the first one. This I thought was due to avoidance of certain types of food that was supposed to contribute to their formation; and drinking plenty of water, which helps flush out the kidneys, stopping them forming so big. This second stone came out quicker, passing it the following day. After this one I was sent to a kidney stone clinic, to find out what specifically was causing them and received regular checkups and a program of things I needed to do to avoid them forming again. It had been the build up of these kidney stones that had been causing the pains in my kidneys three and a half years earlier, and at the time I had thought it was something to do with the stomach problem.

Anyway, I was still attending the Day Hospital. And in some ways it was probably the best thing for me; due simply to the fact that my health problems, namely eyes and ears had left me isolated and more dependent on the Day Hospital for help. I didn't really know anyone else, and in the state of health I was in it was hard to make friends. At least at the Day Hospital there were people to talk to and help when I needed it. My medication, the sulpiride, had been increased also, and there did seem to be a lessoning of the so called psychotic type symptoms that had troubled me.

Fast forward six months: now January 99. And there had been virtually no progress with the white noise generators. And I had only been able to withstand them in my ears each day for 15 or 20 minutes at the most,' then would have to take them out because my ears would start to hurt -- to the point of being unbearable. And they had made little difference to the worst aspects of this problem. And I was still in the same situation; the only difference was that I had adapted to it, and learnt to live with it. One example of this: finding cafes or restaurants' where no radio was on, or music was being played. And there were a few of them scattered around north and west London; about five I'd found altogether, and which I would often go to for a meal, and often to pass time.

My world had changed dramatically. No longer able to use my eyes in the way I used to; no longer able to read anything whether for pleasure or necessity. And now with my ears being in this state, where I could no longer listen to anything that came through a speaker, even on low. The outcome of this was to isolate me even more; cutting me off from all kinds of things I had so long taken for granted. And now was left in a state where for most of my time all I could do was think and observe. I had become the thinker and the observer. And instead of participating in the world, overwhelmingly, I was observing it from the outside. And where before, I had been able to listen to the radio in my flat that pleasure was no more, and now went home to a silent flat to ponder and think about life and my own situation in the deafening silence that my flat had now become. And that wasn't all, there was more bad news. I'd been having a recurring problem with my neck, which a few months earlier I had somehow sprained while awkwardly turning over on my bed one afternoon while resting: this is what it seemed like anyway. It started off as a sprain and I was given anti-inflammatory tablets by my doctor to help heal it; and had had difficulty turning my head to the left because of it. And this now, had somehow spread to both sides of my neck, and I was now having trouble turning my head both ways; it was very stiff all the time. I could only turn it so much, and then it wouldn't turn any further.

I'd had it x-rayed. There was no injury to any of the bones, so the injury I was told was of a muscular nature. I was also sent for physiotherapy, but this didn't make any difference; and exercises I was given to create more movement in my neck sometimes had the opposite effect. By encouraging me to try to turn my head that little bit further they had instead caused more sprains, and I had had to take tablets again to bring the swelling down. It was a vicious circle. And what was also gradually happening, was my ability to move my head up and down was also becoming more restricted. What a mess I was in!

Depression was now a large part of my life, which I lived with and managed the weekly ups and downs of, all the result of my never ending health problems, and the frustration, restriction, and isolation they had brought. And the not knowing if my life would ever be normal again. Some bouts of depression at times were so bad that I couldn't get out of bed. I remember one morning in particular I felt so depressed I literally couldn't move, it was like I was frozen, like I had been drained of all energy. Anyone who has ever suffered with bad depression will know what I'm talking about; as at this stage it's a more clinical type of depression. And anti-depressants can only do so much. If you have an ongoing scenario that's causing the depression then it's not going to completely go away. Like the other health problems, I had learnt to live with it and find ways of dealing with it. I knew that staying in bed when it was bad didn't help, and if I got up and moved around a bit I felt better; or going for a walk would lift my mood. My despair, frustration, depression and anger at times was such that I had come to that point where if I had had a gun in my possession I would have put it to my head and said, fuck it, bang!

One positive thing throughout all of this is that I had not resorted back to being a smoker. And through all of this I had managed to stay off what had been on average forty cigarettes a day habit; and by this time had not had a cigarette since 1993 – over five years.

I still continued to look outside of conventional medicine in the hope I might find cures for the ailments I was afflicted with. And in March of this year went to see what is referred to as a 'Psychic Surgeon', a man in Chelmsford called Steven Turoff, who apparently people came from all over the world to see. He had cured many people of many different things, and there had been a write-up about him in either the September or October edition of cosmopolitan magazine the previous year, which someone at the day hospital had read to me.

I had heard about him before this. Someone had mentioned him on the radio one night not long before my ears had gone bad and I had made a mental note of it. But it was the article in the magazine that had persuaded me to go and see him.

But although he had an impressive percentage of cures and literally loads of letters on the walls of his waiting area from people he had healed of all manner of afflictions; on the day I saw him he did not cure any of mine. But I didn't give up there and then, he told me to come back again. And I knew that sometimes things were not always cured straight away, so I made another appointment for a later date. The fact that he had a long waiting list and had cured a lot of people; and people came from around the world to see him was impressive enough for me to go back a few more times. I had nothing to lose by trying, only the cost, which wasn't a lot.

The term 'Psychic Surgery' sounds quite drastic. But there was actually no surgery as we would understand it being done, not on me anyway. This seemed like another form of spiritual healing but using different methods. It had said in the magazine article, that he worked with the spirit of an Austrian surgeon, who had lived during the time of the First World War: and for reasons of a spiritual nature wanted to carry on working on earth, through someone, and chose Steven Turoff. And that it was the spirit of the Austrian surgeon who worked through him doing the work. And based on the number of healings he had done, it was obviously done using methods of a spiritual nature; that we would have no concept of.

As well as the things I had wrong with me, I had additionally to deal with the often constant boredom of being in my flat, with only my thoughts for company. And felt that I needed or had to do something to relieve this, something to occupy my mind for some of the time, something to do, or something to look forward to at least. I couldn't keep this up.

I had thought about what I could do, and had thought about the idea I'd had for a screenplay, first thought up almost five years earlier in 1994. And although my eyes were still in a bad way, I had by now cultivated ways of being able to write down a couple of words at a time by looking at something for one or two seconds, then quickly looking away before the pains started. Even for one or two seconds there was pain; but it was bearable and would soon go away. But if I tried to keep them focused any longer the pains were more severe and would take a lot longer to go. I now knew what my eyes could take, and what was bad for them. I had discovered this when having to sign my signature for various things. As a signature could only be signed by me, I'd had no choice but to force myself through the pain of trying to focus. And 'signing in' was something I had to do every time I went to the Day Hospital. And it was hard, as the smaller something was the harder it was to focus on. I gradually got better at doing this, and I now thought that if I write in larger writing maybe I could start writing this screenplay; at least have a try at it anyway.

I along with two other people had actually written a play when I was in the Drama Production Group. The last play that the drama teacher wanted to put on had to be one that the whole class had a role in. There were twelve in the class, and we were having trouble finding a suitable play, and I along with these two had come up with an idea for a play which everyone would have a part in. And we set about writing it. The nearest description was that it was like a 'Hammer Horror' in a modern day setting. The drama teacher didn't like it and by the time it was finished she had found a suitable play anyway.

I had done the bulk of the work on the play and put the finishing touches to it. And although disappointed that she didn't want to use it, I thought it had potential as a low budget Film; and planned at a later date to develop it and convert it into a screenplay. I had also decided at that time during that early summer of 1994 when everything was going so well that I would look into filmmaking and directing courses too. And by that September I had also started doing a directing course one evening a week. My eye problems had started during the later part of that summer, but at this stage were not that bad. The directing course was cancelled after only six weeks due to a lack of people, but what directing I did do during the course I really liked and was good at; and felt in my element. And knew it was something I could be very good at. And it was not long after this, while listening to the radio one night that I had had an idea for another screenplay. And as the eye problem progressed, and I was no longer able to work, and had a lot of time to think, I'd decided once my eyes got better I would start writing this one first. And that instead of just trying to pursue an acting career, and all that that entails, that, I would work towards making my own films; and act in them also: and had lined up a filmmaking course that would get me started. It was something I could put my energy into instead of just waiting around trying to make it as an actor, which at the time due to the British film industry not making many films, acting jobs were not in great supply. I was confident of my ability, and knew I was very capable of making my own films "and starring in them." And which, I planned to start off by making the low budget converted play' into a film. This would be my first film. All in all I had also felt for the first time like I had finally found a vocation. And that this is what I really wanted to do. But little did I know, what the future had in store for me; and now nearly five years later, here I was, contemplating writing this screenplay; but for very different reasons.

I'd often thought about it over these years adding to it in my mind and dreaming that one day I would get it written, and make it into a film. One good thing about this time span was that from the initial idea it had over these years grown and expanded, and as well as the main plot, I now had the beginning middle and end; and a name for it. It was called 'The Laughing Detective.' A 'murder mystery with a twist: and elements of the paranormal. And it was one night not long after my decision to at least have a try at it anyway, that I sat down and began to write it: and gave myself a starring role. And on the night over the space of a few hours, managed to write seven lines of dialogue of the first scene. And it made me feel very good; more so because I was actually doing something. The sense of worthlessness that can come with being incapacitated had suddenly been reduced! And finally, but more importantly, like lights, camera, action, I had begun to write 'The Laughing Detective'.

The year was moving along, and progress on all my health problems except for my legs was almost none existent. My legs had come a long way, going from strength to strength. And during the previous year, special insoles, to fit inside my shoes had been made for my feet: in order to raise a part of the foot in a way, that would align my legs properly; being told that my feet were flat and that this was part of the overall problem. And had been sent by the National Health Service, to have them specially moulded, to fit my feet. And these were now contributing to this progress.

I had long ago left behind the physiotherapist. I knew all the exercises I needed to do to get them right, and had done them on my own, at my own pace. And the fact that I was now walking unaided by a walking stick, and walking good distances without needing to stop, except occasionally; this itself was making them stronger. And I could now walk up steps without needing the walking stick. I still needed it for some things, but now had a fold up one instead, which I carried with me and would use when needed.

This was some good news! And one thing, that was getting better. There had been no improvement in my neck, in fact it had actually got worse, and I had turned to alternative treatments because of this: One of them being Osteopathy, and had five sessions of it. This did help a bit; it created more movement for a few hours, or a day. But soon after, would resort back to tightening up and being stiff. And I gradually realized that it wasn't going to solve the problem and stopped going. I was sure that if it was going to make a big difference it would have done so before five sessions. But not long after this I started a similar form of treatment called 'Cranial Sacral'. And as well as it being for my neck I was told by the Practitioner it could help solve the problem with my ears as well; as he explained: by rebalancing the rhythms - as it also concentrated on the head area. I made regular appointments once a week or fortnight and stuck with it for a period of four months. But this too had a similar outcome. But I stuck with it longer because of my ears; in the hope that it would improve the hypersensitivity. But it was towards the fifth month, and after virtually no real improvement that I started to get really fed up with it; and he suggested I try a Homeopathic remedy to compliment the treatment. I did so, which he prescribed; but it had an adverse effect, causing even more stiffness in my neck -- and now around my shoulders too; even making my upper spine between the shoulders stiff and feeling extremely sore. It was a severe reaction, and it didn't fully go away; and which left me feeling very angry. And while he was giving his reasons and excuses for the bad reaction, saying that it may have been because the dose was too high; I just felt conned and even angrier; and feelings of smashing a chair over his head had fluctuated in and out of my mind. And now, finally realizing that this Cranial Sacral wasn't going to improve anything: except the bank balance of the Practitioner, I cut my losses and moved on; leaving there worse than when I started. Another expensive load of new age baloney was my conclusion of it.

This further stiffness and soreness I was now experiencing in the spinal area, between the shoulders, because of this, had to a much lesser degree already been there, and had had me feeling at times that it was some type of nerve damage and that may well have been responsible for stopping the neck muscles moving and working properly. But I really didn't know for sure where it had come from, or what it was? But had seemed connected with the overall problem. But now, it was there constantly, and much worse; and now, something more I had to cope with.

I had gradually grown more cynical about the alternative health profession. Also, it had now been about 18 months since I started having the spiritual healing, and was in no better state of health for it. And by now, had also been back to see Steven Turoff three more times. But psychic surgery and healing he had done on me had achieved nothing; and after the last visit decided I wouldn't go back again. I'd seen just about everybody in the book for one ailment or another, and had noticed many of the various treatments and Practitioners in this field promoted cures for just about anything; whether using Herbal medicine to Acupuncture, Homeopathy to Osteopathy, and so on. But none of it had worked on me or cured any of my health problems; and was now feeling that a lot of it was just quackery.

Another thing bothering me was the hearing ailment 'the Hyperacusis' and lack of progress; virtually none at all. And from the information I had now gathered, it seemed like I had a much worse case of it than was normal; in fact, a very severe case of it. I was still using the white noise generators but now in a limited capacity, the hearing therapist now telling me because of their effect on my ears to use them for just 5 or 10 minutes twice a day or whenever I could. But I knew other people could use them for a couple of hours, and then eventually even longer. And my five or ten minutes here and there, my only being able to use them in this limited way, I had to be honest and think, at this rate, they were never going to get better. And this is why in desperation for a cure I had also looked for alternative solutions or cures.

One good thing that was going well was my screenplay. It had now been four months since I started it, and I had completed fifteen pages of script, and made lots of notes. And I found that my ability to focus and write had greatly improved. This was due in part to a special pair of glasses being made up for me, with an effect in the lenses called a prism. Its purpose was to force the eyes to focus better; which they did -- having a beneficial effect by enabling me to focus on things for longer. This unexpected help had come about by going to an optician's for a sight test six months earlier. I wasn't even looking for any help with my eyes, but just finding out if my sight had further deteriorated. It was the optician who had mentioned these prisms, and had referred me to a university where opticians are trained (and where the prism prescription was put together) after noticing, what an ordeal it was for me trying to focus sufficiently enough to complete the eye test. I had now been wearing these specially made up glasses for three months, and could now focus and write for about five or six seconds; and there was certainly more strength in the eye muscles, and this was good news. And another improvement was I'd gradually got rid of the dark tints. I no longer needed to use them. The photophobia was still present, but now to a lesser degree. I had also been using a natural system for improving eyesight and eye problems I'd found out about called, 'The Bates Method'. This was comprised of some gentle eye exercises (nothing to do with using a pen) and a technique called 'palming'. And another part of this Bates Method combined the gradual reducing of the prescription strength of the glasses; which I had now done as well. The palming was the placing of the palms of the hands over the eyes, totally covering them, blocking out all the light, for five, ten, fifteen minutes, sometimes longer. This was usually done while sat at a table with the elbows on the table for support. This palming brought about a deep relaxation of the eye muscles. And it was after the palming that you were supposed to do the eye exercises. The Bates Method theory was that tensions in the eye muscles is what caused vision to alter or other abnormalities to develop. I had been having appointments with a Bates Method Practitioner who had been teaching me all the exercises and methods. It was early days yet, but what I had found the most helpful was the palming, using it frequently, usually after a strain brought about by trying to focus on something for longer than my eyes could withstand; finding that the pains would go away much quicker by just doing the palming for a couple of minutes. And I had started to use this when I was writing the screenplay, enabling me to do more. And during the summer I did a screenwriting course, one evening a week for six weeks, feeling confident enough, knowing, that I could now write notes. The only problem encountered was when demonstrations were shown on a television and I had to put my fingers in my ears, as well look away, as the brightness of a television screen was also' still a problem. It was a good course, and gave structure to the whole process of how a screenplay is written: which is basically a formula that overwhelmingly nearly all films are based on. I now had all the information I needed; from the structure, to how many pages in general a script should be, to the correct terminology etc. It was also during this summer that I had been assigned a Social Worker. To help me, with the various problems I encountered due to my situation.

It had now been close to four years since I was first admitted into the Psychiatric Unit; and I was still attending the Day Hospital. My still being here was due in part to a combination of both physical and mental illness, and the vicious cycle of how the physical health problems often affected my mental state. And with the schizophrenia I was being treated for, there had been another raise in the dose of medication I was taking for it.

I had got use to the routine of the Psychiatric Unit, and the haven that the Day Hospital had become for me. And over the time I had spent here both on the ward and Day Hospital I had seen many people come and go. And had seen people with all manner of different mental illnesses, as well as the problems that result amongst patients; from altercations to arguments, random acts of vandalism, or patients having to be overpowered by several members of staff, and given an injection to calm them down; or just the general bizarre behaviour of some patients. All in all I'd grown used to it, and by now when any incidents occurred, didn't blink an eye. In fact some of them were very amusing and would easily pass off as entertainment, and this is how I now often perceived it – as a bit of entertainment. And although a place with a lot of misery; there was humour there too.

It was in September of 1999, during a weekly meeting with my care coordinator, that I had discussed the gradual realization that I had been mentally ill for years, and hadn't even known it. This gradual realization had occurred over the previous six months, and I had elaborated further on this with the Psychiatrist --- obviously the medication had put something right! And as I looked back over my life I had reached the conclusion that this schizophrenia had probably started when I was 19. And throughout the years had manifested itself in many ways; delusion, paranoia, minor hallucinations etc, as well as the effects on my behaviour that this had produced. I'd looked back on many incidents over these years, and behaviour patterns, and realized I had been clearly ill. This gradual realization was horrible; and somewhat sad also. I had thought when first told of this diagnosis a couple of years earlier that I probably had a mild case of this schizophrenia. But now realized it had been much more than that; as my 'care-coordinator remarked', when ill with this, you're the last person to know it. He was right! And it was not long after this, that I think the process of discharging me was put in motion.

A couple of months earlier than this, I had decided I would not bother with the spiritual healing anymore. I had given it long enough, and now just wanted it out of my life. It may have worked for other people, but after two years of having this healing from different people, and at various places, and sometimes popping into other places run by the National Federation of Spiritual Healers, for additional sessions: as well as trying a Japanese form of spiritual healing called Reiki; I had now finally concluded that it wasn't going to work on me, and had decided to move on from it. However, it was shortly after making this decision, about a month later while visiting another Osteopathy Practitioner, because of the worsening state of my neck, which by now the stiffening of the muscles had spread down to my upper back, and was now also restricting my upper body movement. That in desperation I decided to try osteopathy again.

The osteopathy I'd had before had helped a bit, so in this desperation because of this worsening situation I had decided to try it again; but with a different practitioner. And while there when asked about any other health problems, I mentioned my other ailments. And towards the end of the session, the practitioner told me about a 'Spiritual Healer she knew', and had seen herself, saying: that she had powerful hands, and handed me her business card. By now, I had given up any inclination that spiritual healing would work on me, also feeling that I had moved on from spiritual healing; and not really wanting to get into a conversation about it, just politely took the card.

The couple of weeks following this, I thought about what she said about this healer having powerful hands. And it played on my mind a lot; and had made me rethink my position. And, what I had learnt about spiritual healers' was that some were more powerful than others. And the reason for this was that apparently some people were a better channel for the energy. And I suppose, out of desperation and the will to get better, and thinking that maybe she was a more powerful healer, I decided to give her a call; and by now, September, had already had two appointments with her. She was a lady in her fifty's, called Margaret Palmer.

It was my neck and upper back that were now causing me the most discomfort and stress, as well as pain. And when I called her, she said she had had good results with backs. And it was this also that convinced me to go and see her. And I knew as before, that it would probably take several sessions before I would see any results, if any at all.

I only went back to see the Osteopathy Practitioner one more time. It seemed that my neck and back had now become so much worse, to the degree that the osteopathy now had little effect on it, and decided instead to opt for the spiritual healing. One good thing that I had on my side was that, due to my various health problems, I had been categorized as disabled and was given more money by the Department of Social Security because of it. This had allowed me to at least try to improve things, and also seek cures by alternative means. And also help myself in other ways, which I was grateful for. It would have been hard going and a double blow to have been impoverished, as well as incapable of being able to earn a living, and have to survive on basic social security; which isn't a lot. And in my situation would not have gone far.

Another thing I had been doing, was trying to get rid of the strapping tape that kept my kneecaps in place. I'd thought my legs were now strong enough that they no longer needed it, and had made three attempts to get by without it. Though each time, after walking about a hundred yards, the kneecaps would start to slide out of place, and I would have to tape them back up. It didn't seem like I was going to be able to manage without the tape, and had resigned myself to that reality. Feeling that there must have been some permanent damage that was irreversible; it couldn't have been anything else really. It would have been nice, to not have to go through the routine of having to strap them up every day. But at least I could now walk – and that was the most important thing.

A doctor had told me, two years earlier, about an operation that could be done on the knees to keep the kneecaps in place; but said that it didn't always work, and would sometimes make the problem worse. And because of the risk of it not working, or further damage, decided it wasn't a risk worth taking.

The situation I was in was difficult enough, and I had enough to cope with. But a development of a different nature had recently presented itself, i.e. the neighbours in the flat below me had become very troublesome. They were a retired couple, the woman was ok; it was her husband who was the problem. He had developed a habit of going to the pub almost every day, then coming home in the early evening very drunk; then often getting into arguments with his wife, which would usually last a couple of hours or more. This would include doors being slammed, and things being thrown around; and the volume of the television going up and down, often excessively loud. Some of the noises were unbearable to my ears, mainly the television and the door slamming. And the sporadic unpredictability was just as stressful; having to put my fingers in my ears on and off while this was going on. I had spoken to him about this when he was sober and he would apologise etc, then when he was drunk it would happen all over again. Some nights it was that bad I would have to go out for a few hours, often just riding around on buses to get away from it. But mainly to stop my ears from flaring up, as a bad flare up would make the sensitivity even worse, and would often take a day or two for them to recover or calm down again. Usually by eleven o'clock he had gone to bed and I would come back.

This situation had slowly developed over the previous four months, and gradually became more frequent; now almost every night, to the point where I wondered when I came home or heard him coming in, what kind of a night it was going to be tonight. The neighbour in the first floor flat below them eventually made a complaint to the Housing Association, and so did I, and he was visited by someone. This helped, as after the visit and repeated warnings the disturbances became less frequent. And I think this was also due in-part to his failing health, as he had a respiratory illness, either Asthma or something else. What I did know was that on days when this illness was worse he would have to use a machine called a 'Nebuliser' to help him breathe, and this had recently become a big problem for me. This because of the noise it made. It sounded similar to a washing machine. But a much more piercing denser sound, and louder, and which had a pattern to it. It would make this sound for about ten seconds then stop for about four seconds, then repeat this again. Not only was it aggravating but it was also having a detrimental effect on my ears. Any type of electrical sound, whether something coming through a speaker or these dense machine type noises had the same effect on my ears, making them ring, causing burning pains and throbbing, which if I didn't soon move myself away from would gradually become more severe, causing a flare up.

He obviously needed the machine. And in normal circumstances I could have just turned on the television or some music, and this would have helped to drown it out. But I was still unable to do that, and had to put up with the noise of this machine coming through the floor. The only place where I couldn't hear it was the kitchen; and would often find myself sitting in there to get away from it. I didn't mind this too much as most nights I would now work on my screenplay. Sitting at the table in my dimly lit kitchen, which faced a window, and often gave me an inspirational view of the day or night sky. I had found ways of dealing with it. But now what made the situation worse was he now kept it on when he went to bed, all night, almost every night, whereas before he would just switch it on for periods of time, throughout the day and evening while awake; and it was now stopping me from sleeping. I tried wearing earplugs while I slept, but this dense noise would gradually pierce through, causing the horrible symptoms.

He kept this machine in his hallway right next to his bedroom – which was directly below my bedroom. And it became that much of a problem that eventually most nights I found myself having to sleep in the kitchen, and not even on the kitchen floor. But, because my neck and now back, was increasingly becoming even worse, I was unable to bend adequately enough to be able to get on to the floor; it had now literally become that bad: now to the point that the top half of my body was almost frozen stiff. And as well as now additionally not being able to hardly move my head up or down, or turn it side to side, I couldn't twist or turn my upper body either! It now felt as stiff as a board. And although I had no longer needed the walking stick for my legs, I now needed it again to assist my getting in and out of chairs, as I couldn't do this without it – I couldn't bend my back sufficiently. It seemed hard to believe that what had started off as a slight sprain in my neck had now turned into this. And what I ended up doing was pushing the kitchen table up to and in line with the washing machine and putting cushions and a quilt over them, a makeshift bed; and not a very comfortable one.

Some nights when the nebuliser machine wasn't on and I slept in my bed it would sometimes come on in the middle of the night, and I would be woken up with screaming throbbing ears and would have to transport myself back to the kitchen; and not always able to get back to sleep, often being deprived of a decent night's sleep. It seemed like every area of my life was aggravation. And now I couldn't even guarantee I would get a proper night's sleep either. That had now become unpredictable; which I really resented, and became increasingly angry about, sometimes wishing he would hurry up and die, or have a fatal accident. This is what my thinking had been driven to. The man in the flat below him had expressed something similar: Angered by the constant drunken disturbances had one day said to me, "I wish he would walk in front of a Bus". This was at a time when I was trying to make more progress with the white noise generators. But found the aggravation of this machine, combined with the noise of the still occurring drunken disturbances, were often enflaming my ears. And trying to use the white noise generators during these times when in an enflamed state was almost impossible. And usually had to wait a day or two for them to calm down or fully recover. And this enflaming of my ears was happening almost every other day. I couldn't keep on living in this situation. It was driving round the bend. At times I felt like walking in front of a Bus myself. And with this latest development' there was no way my ears would ever get better. Just when I had made improvements in other areas along comes something else. At times, I thought, as far as misery and Hell goes, I'd had the book thrown at me. And it still wasn't over. The will to get better and do something with my life was as strong as ever; but at times I thought something was working against me. And something I had no control over. So much so, I had even visited a clairvoyant' looking for answers.

Anyway, my social worker had contacted the Housing Association about this Nebuliser machine and had explained the situation; and was told about a soundproofing programme they had, and said I would be put on the list, which was usually a year to eighteen month wait. In the present circumstances I didn't know if I could stay in this situation for that long, and had thought about moving, and even going back to Wales; renting a small house there.

The Housing Association also had a mutual exchange programme that allowed tenants to swap flats with each other, or with other housing associations. But the trouble with moving to another flat was that I might encounter a whole new set of problems that could be bad in other ways. It wasn't an easy option, and I needed to be somewhere quiet; and started to seriously think about finding a place in Wales, as it seemed like the safest and most predictable option. And I would have help from my family. And also over the previous few years had been going there more frequently anyway; mainly due to my situation. And would stay in a small caravan my parents had in their garden. And a couple of years earlier my father had even got me a seat on the local community council, which he had also been elected to -- won in the local elections; which met once a month and which sometimes I would have to turn up to. At that time there was no pay of any kind for this particular body. But for many was generally a stepping stone to becoming a 'county councillor', and which my father later became, as well as the mayor. Personally, at this particular time it wasn't something I was interested in; and because of my situation, even really cared for. But having a seat and a vote helped my father out, and I had gone along with it for this reason. It had been his idea to put my name down on the ballot, and on the back of him I had got elected also, taking a seat from one of his rivals: and served one term.

1999 was now coming to an end. And I had spent that Christmas in Wales. And, had seen in the New Year' and the beginning of a new millennium. I was glad to see the back of the 1990s. For me it had been a decade of hardship and suffering; that still hadn't ended. I only hoped the next decade would be kinder.

Back in London: and now January 2000. It was in January, that I received confirmation that I would be discharged from the Day Hospital within the next three months. And a month later was given the date of discharge. From what I remember it was the first week of May. Mentally I felt quite good; and was somewhat glad I was being discharged. But also uncertain of the future. And mainly because' I was still in a physical mess and couldn't go and do the everyday things that most people took for granted. 'My life was still going to be hard'. And just over three months later, when the day arrived; I was finally discharged.

In the weeks leading up to this, a care in the community plan had been put together for me. This would include a weekly visit to or from a social worker: to things I needed help with. And also having, initially, to see a Psychiatrist every four months, who was based at the Care in the Community Office, along with my social worker – key worker.

It had been four and a half years, since I was first admitted into the Psychiatric Unit, and on the day it felt strange to finally be leaving. And when the time came, I had a last look around, said goodbye to certain people, and off I went -- the end of a chapter, and the beginning of another.

The struggle or battle with my physical health continued. And despite seeing this new spiritual healer, my neck and back had not improved, but continued to worsen. I had even seen a specialist consultant' at a hospital, which dealt in this area, and had also been sent for a course of Hydrotherapy (physiotherapy while in a pool of warm water). But none of these different medical type physiotherapy treatments I'd had for this, had made any real difference. Some seemed to help or slightly ease something for a few hours, then, everything would tighten back up again. The state of my neck and back was now at a point where this stiffening had progressively spread all the way down my back affecting and hardening all the back muscles. And this itself, had over time created a weakening of the stomach muscles; through their lack of not being used. They were getting no exercise, and which I found would now rupture very easily if I tried to lift or move things. And these weren't necessarily things that were heavy, just everyday items – a kettle of water or a light bag of shopping. I had to be very careful, as I'd already caused ruptures on a few occasions. One area on the lower left side of my stomach seemed to be particularly vulnerable, and was the first area to rupture and seemed to be the weakest point. And although the ruptures would eventually heal, this area never seemed to completely recover. It was almost always sore or very sensitive; almost like an open sore, and would now have to place my hand against this area when lifting or moving things. I also had to watch how much I ate, as a full stomach would stretch or push against this area, further aggravating it. As far as my neck and back were concerned, I remember one of the physiotherapists I'd seen, saying, that there was absolutely no reason why the back and neck muscles were the way they were, saying "there was no reason for it". Their obviously was. But it seemed to have eluded everyone.

It was now June 2000. It had been a month since I had left the Day Hospital. The summer was now approaching and my increasing frustration with my worsening back had been really getting to me. One night in my flat, I had become so frustrated and angry with it, I'd drank a whole bottle of red wine, then decided to do some physiotherapy exercises, to try to get some movement into the muscles. Although I had been not much of a drinker; during the last two years, I'd developed a habit of often having a glass of red wine during the evening; and usually looked forward to it. It had the effect of elevating my mood, and making me forget about the hardship; as well as easing some of the physical pain. But as my situation became worse, I sometimes found myself finishing off a whole bottle.

Anyway, on this particular night, I was feeling the effects of the wine, and didn't even care if I further injured myself. I was fed up and frustrated with the state my back was now in. The whole top half of my body was now frozen. And I just went mad with the exercises, not caring what happened, anything to get some movement, and was really pushing and forcing my back and neck into all manner of positions. And fuelled by rage, anger, and drunkenness, and the red wine acting as a pain killer, I managed to loosen up the muscles quite a bit; and not long after fell asleep, that night sleeping in my bed. On waking the next day, my back felt raw and stiffer than ever. And I soon realized that I had further injured it, by the pain that was now coming from the top half of my spine. So bad, that I went to the accident and emergency department at the Whittington Hospital. I was x-rayed and told there was no injury, but that I had arthritis in my spine, right where the pain was coming from, and showed me where on the x-ray. Concentrated between the shoulder blades, and right where the shoulders meet the neck, and further down the spine.

As I had been experiencing pain in this area for quite some time I had assumed it was tied in with the overall problem. And although there were no injuries, my angry attempt to improve my back and neck, had aggravated what was now found to be arthritis; and my back now felt in a worse state than ever. And being told I had arthritis was more bad news, which just sent me into a depression.

At times I think the only thing now keeping me going was my screenplay, and my desire to finish it; which I was now halfway through. .And had thrived on the creative process involved; the story and characters unfolding; and the incorporating of my own soundtrack into it. This at the very least' was giving me a lifeline. 'I still continued seeing the spiritual healer Margaret Palmer'. I didn't think it was going to make my back and neck better, and had some months before told her this and said that I didn't want to bother with the spiritual healing anymore. She had encouraged me to keep having the healing, and told me to have a think about it. And a few weeks later I went back. It was also a source of support,' and I did in some way always feel uplifted after. And I had noticed that it had sometimes diminished pain and eased some of the bad symptoms, particularly helping with the ruptures – which I thought was encouraging, thinking that none of the healers' I'd seen before had had these effects. And maybe in time it might even heal me. And for those reasons I went back.

July 2000: And although now in an increased incapacitated state and increasingly becoming more depressed because of it. And still suffering the effects of my neighbour and his Nebulizer machine; around this time I had been arranging to move back to Wales. And with some help from my family, had been lining up a place to move to. Although I didn't particularly want to move and was in no fit state to move; for my sanity and overall wellbeing and to get my ears better and improve other things, I felt I had no choice, and had planned to move in September.

Chapter Five

The

Magical Whirlwind

A month later, now August 2000: The situation with my back had now become very serious. It felt like the additional presence of arthritis, that had showed up on the x-ray had accelerated, and was now spreading unusually quick, moving further down my spine and further out across the shoulders, and, my whole back and neck, now felt like it had ceased up. And the stomach muscles' growing ever weaker were now very easily ruptured, even more so than before; and, I was having to deal with a lot of pain and discomfort from both of these. I went back to my doctor, to see if anything could be done and was sent to the hospital for another x-ray, which I knew was a waste of time, and only showed that the arthritis had got worse, and spread even further down my spine. The doctor, also arranged for me to see a Physiotherapist, to help and try to strengthen the stomach muscles.

As far as my back and neck was concerned, I'd done everything and seen everyone in this field that I thought could help me, there was nothing more I could do. By the second week of August a week later, the upper part of my spine had become so bad that it now felt like it had turned to stone. And the upper half of my body would now sway about like it was on a spring. And this would affect my balance when I was walking. I couldn't understand what was happening to me; and was powerless to do anything about it. I was in a desperate state, and for the first time felt uncertain and scared. I felt like the end was near, and felt close to death; like something was closing in on me. But strangely enough' there was a part of me that now didn't care either. I was tired of fighting and battling, I'd had enough, and was at the end of my tether. I had battled and adapted to all the other health problems, but this one seemed out of my control, there was nothing more I could do. It felt like my upper body was now collapsing, and at some point, and very soon, something had to give.

I'd been thinking about what I'd been through, and was still going through. I had persevered and persevered, and I now felt I could take no more. And in this deepening despair I had quietly referred to God,' saying, I wish you would just hurry up and take me out of this hell, I've had enough; I can't take anymore. The will to try and get better; if Olympic medals were given for trying and sheer perseverance, I would have won Gold. I'd thought about what I had been through over these last number of years, and it had been a voyage through hell. And with the current situation, I saw no way out. It was now out of my control. I'd been through every emotion in the book, over and over and back again and had had enough. I was tired and drained and worn down by it all. I felt like I had been on an endurance test.

But it was one morning, during this second week of August, during this very bleak escalation and worsening crisis; that I had felt that something was coming to a head. And it was on this particular morning and hot summers day that I had left my flat around mid-day, to go to the hospital because of an even severer development of additional symptoms; and had walked along Tufnell Park Road up towards Tufnell Park; which is about two thirds of a mile walk, and slightly uphill; and where I intended to wait for a bus. It was a glorious hot summer's day, the height of summer. And as I slowly walked along the road, tired; in a lot of pain, and feeling faint and lightheaded; and with the added pressure of the hot sun beaming down on me. And, on gradually reaching the steepest part of the road, about a hundred feet from the end, a real feeling of exhaustion, came over me. I felt terrible and weak. I walked a bit further, and now, feeling fainter and even weaker. And suddenly feeling as though I was about to lose control of myself. And it was at this moment that I fell forward, about to collapse and fall flat on my face. And to weak and powerless to stop myself from falling, and not even trying to fight it. I just thought to myself, fuck it' and felt like I was about to die. And as I fell forward, accompanied by a sense of losing consciousness; all of a sudden a powerful swirling mass of energy surrounded and engulfed me, stopped my falling, propped me back up, took control of me, and marched me up the hill. And as it did so, this swirling mass of energy seemed to be pumping into every part of my body; literally enmeshing me in it; swirling, pumping, surging into every part of my body, my legs and up through my arms and swirling all around me and all over me; it was like a mini whirlwind, and continued until I'd been brought to the top of the road; then gradually tapered off, and seemed to disappear.

Bemused by what had just happened, and feeling a bit better, I paused, and stood there momentarily; then I crossed the road. And as I walked, I again felt the presence of this energy still swirling about me. But now much more gentle; and nowhere near as powerful. I walked a bit further, and sat down at a bus stop; and just sat there, not stunned, or amazed, but just very calm, tranquil, and feeling numb; and knowing that something had just helped me; and something that was obviously otherworldly. And gradually, realizing' that something must have been looking out for me, and may have saved me from death; or at least falling flat on my face and serious injury.

I sat there for a while, taking in what I had just experienced; and in the normal scheme of things I suppose I should have been amazed. But I wasn't! Experiencing something this amazing when you feel close to death doesn't have the same effect, or sense of awe as it would under normal circumstances, and would be safe to say that this would probably be the same for most people. I was severely depressed and had felt like I was about to die, I'd felt terrible, and deathly ill; that's how I felt. But after sitting at the bus stop for a while, gradually I started to feel much better. And although I had been on my way to the hospital, I now felt like I didn't need to go there, and instead decided to go to the Waterlow Park in Highgate and just sit in the park for a while. And what I noticed as soon as I stood up to get on a bus, was that this energy which I thought had now gone, was still with me; swirling and swishing about around me. I could feel it gently swirling around my arms, and could feel it being pumped up them through the palms of my hands. And after I had got the bus to the park, and after getting off the bus, as I was walking, this energy was still with me the whole time, swirling about. And even when I was sat on the park bench it came and went several times. And over the following two days this became a regular pattern; this swirling energy would sporadically appear: usually when I was outside walking, but even when I was on a bus it appeared and would swirl around my arms and pump up through them into my body. Again, this wasn't as intense and as overwhelming as the initial whirlwind type energy had been, but a much lesser degree of it; and measured. And it didn't completely surround me, but was strong, and would sometimes make my arms flail or swing about. And what I also noticed was that I started feeling a lot better and my back began to feel a bit better too; a noticeable improvement and slight easing of the worst symptoms had occurred.

It was an unusual phenomenon. But during the two weeks leading up to this there had been some strange goings on. When I had last been to see Margaret Palmer for spiritual healing; during the healing while sat in the chair, my hands resting on my knees and palms facing upwards, out of nowhere I started to feel swirling balls of energy in my palms. This continued on and off the whole time I was sat in the chair. When it first started I told Margaret what was happening, and she remarked that maybe it was a wakeup call for me to be a healer. I took no notice, thinking only that the state of health I was in, being a healer was the last thing on my mind, it didn't even come into it; and that it was me who was in desperate need of healing. But nevertheless I was quite impressed by what had happened, and it had made me think that there was more to this spiritual healing than I had realized. And this wasn't the only unusual thing that had happened. During these two weeks I had also noticed on several occasions while outside, this same invisible swirling energy that was now sporadically appearing throughout the day; I had noticed this too! However, it was much gentler than anything I was experiencing now; and not as noticeable. At the time first thinking it was a breeze or odd wind pattern, or something of that order. But it was unusual enough, and became more noticeable that I had eventually mentioned it to Margaret.

Anyway, it had now been three days since this magical type whirlwind had appeared and rescued me. I'd already had on appointment booked to see Margaret for my regular session of spiritual healing; and booked for this third day. I had a lot to tell her, and while there told her the whole story, what had happened and what was still happening. I don't know what she thought of it, I think she may initially have been sceptical or even just dumbfounded not really knowing what to say, or figure out what it was. She asked me questions about it, and I just more or less repeated what I had already said, what had happened, and was still happening; and said that I was also feeling generally better because of it.

The next day, day four – something different began to occur: And occurring while I was sat on a park bench. The swirling energy appeared, swirling around my arms, then began to pump up through my hands up into my arms. But instead of it just going into my body in general' like it had been doing, it went straight to my spine; right where I had the arthritis. And it soon became obvious that this energy was working on the arthritis, as I could feel it working away at it. And as it did so it became painful. And over the next two days this became a regular occurrence; almost every time I sat down somewhere this process would start up, and as the work in this one area seemed to be progressing: the pain also becoming worse with it.

There was one part of my spine, in the area between the shoulders, and where the shoulders meet the neck, that the arthritis was the worst and most concentrated: and may have even caused the bone to curve or deform. And it was in this area that this healing energy as I now began to call it was mostly working on. It was also this area and top half of my spine that felt like it had ceased up and turned to stone; almost like barnacles had grown over it creating a weld of some kind. And I could only assume by the concentration of this energy in this area: was because it was as I have described it – that severe.

It was on day six of this and during the evening that I phoned Margaret to tell her of this latest development. And that she had said to me that I should ask whatever it was that was helping me: 'that I should ask it what it was?

I was in my flat at the time and it was early evening; and I said yeah' that's a good idea I will do that. And thought when the process starts again I will ask it exactly that. And it was about two hours later, while sat in my kitchen doing some work on my screenplay that this magical swirling energy showed up and converged on me once again; and went straight to my spine, and again concentrating mainly on the same area. This lasted about five minutes and was quite painful. But what happened next was even more weird, and eerie. All of a sudden, as I sat there, still, and feeling like I was being kept in a particular position; my head was slowly turned up towards the ceiling. And I found myself staring at the ceiling, right at the line where the ceiling meets the wall; my head stretched right back, and being kept there, and while kept in this position, that same area of my spine was being worked on. But now something more was being done and used; it now seemed that this healing energy was turning itself into tools or instruments and devices of some kind and were doing what seemed like more advanced and intricate work; like operating on me. And the swirling balls of energy that had appeared during the spiritual healing a couple of weeks earlier reappeared in my palms, and were synchronizing with the movements of these tools, almost like they were powering or manoeuvring them; and they wouldn't just swirl around one way in the palms of my hands, but change directions and move at different speeds – the two seemed to be synchronized. It was another new and bizarre development.

But as I sat there, fixed in this position, and unable to move and sitting and staring at the ceiling, I asked out loud, who are you? And waited for a response of some kind – not knowing what to expect! An eerie calm filled the atmosphere and I waited; but there was no response, no reply. I was kept in this position for about five more minutes; then my head was brought back down to its normal position. I felt relieved. Then a minute later it was turned slowly downward toward the floor, so my chin was touching my chest. It was fixed in this position for a few more minutes, while work was done on the back of my neck. Then it was brought back up to the normal position. And not long after this, my head was then slowly turned all the way to my left shoulder, and then all the way to my right shoulder, and was held in both positions, for short periods while work continued on the back of my neck. This was something new again, another development, and I wondered what was going to happen next.

It was then that I got up out of my chair and went and stood in the middle of the living room. I felt like I had been manoeuvred and guided there; and I was right, because as I stood there, I was then manoeuvred into a certain position; my legs being taken over and moved for me, and my feet being brought close together. And with my feet maintained in this position, my upper body was then slowly twisted all the way round to the left, and held in this position for about a minute, then brought back to the centre. And then after a slight pause was then slowly turned the opposite way, twisted round to the right as far as it could go, and held there for a similar amount of time..

These movements were being done for me, my body had been taken over, I wasn't doing any of it; but at the same time I did feel part of it, it didn't feel alien. And although a part of me was slightly unnerved by what was now happening. I also felt comfortable and at ease with it too. These movements were repeated several more times, then this was followed by other physiotherapy type movements, from twisting, to stretching, and bending type movements. And as these were being done, there was also an additional effect, working simultaneously, synchronizing with these movements. This effect softened or relaxed and eased the tightness and stiffness of the muscles as I was being moved into these positions. I could feel it, a surging liquid like substance surging through the muscles at each stage of the twisting bending or stretching; giving the exact amount of input that was needed to manoeuvre my body. This all lasted about an hour, with the movements gradually getting a bit faster. I then went and sat back in the kitchen; and no sooner had I sat down the movements started again; now with my neck being put through exercises, my head being slowly turned to the left, to the right, up, down, these being repeated over and over. And what was amazing about all of this, was that as well as what I was actually experiencing, this was amazing enough, but also for me was that I hadn't been able to move my head anymore than an inch or two up or down or to the left or right for well over a year. And certainly hadn't been able to turn or twist or bend my body into any of these positions for probably two years.

I don't really remember much else about the evening, or how I felt after, only that it was late in the evening and at some stage I went to bed. But it was the next morning after I had woken up, and while lying in my bed that this process continued: with me now being turned over on to my back. And as I lay in that position, the lower left side of my stomach in the area where the main recurring rupture was -- this was operated on. And again tools of some kind were used. I could feel it being repaired and stitched, the stitching being done by a large needle attached to a bobbin type device, and working very much like a sewing machine, but much slower. This was followed by physiotherapy type stomach exercises being done for me: as I lay on my back my stomach was being pulled in, then out, then sucked in, and held for a number of seconds, then let go. And as it progressed it was sucked in further and held for longer. And as these movements were being done there were pauses in between, while further work was done on the ruptured area. All this lasted about an hour, and then what followed for the rest of the day and evening, was more physiotherapy type loosening up of my upper body. And additionally to this, my legs' were now being put through a number of exercises also.

Although I'd thought my legs were now to almost full strength, this wasn't so, and by the different types of exercises I was being put through I could see and feel where these weaknesses were. And as with my back, my legs hadn't been able to do many of these movements for a number of years. And the worsening state of my back and neck' had eventually stopped any further progress and strength they would have gained through being able to bend etc. And one of these first exercises I was made to do was stand with my feet close together, and then lower myself into a sitting position. I was being made to do this, up and down, up and down, and then as far down as I could go, sometimes into a squatting position. Some of the other leg exercises were more straight forward: working one leg at a time, by stretching one foot forward and the other back, then lowering down my weight on to the front leg, working all the muscles above the knee. I thought at times during some of these exercises that my knees were going to give way, I wasn't sure that they could take all this: even though they were still strapped up. But they didn't buckle or slide out of place; they just got a bit hot. It seemed that whatever it was that was helping me knew exactly what it was doing and knew exactly what each part of my body could take. And after this latest episode, I now trusted that this was the case; and apart from some breaks in-between and going out to get something to eat, this whole process carried on all day and late into the night. Either it was my back and upper body being loosened up and exercised, or my legs being built up, alternating between the two. Then at some point late at night I went to bed. Then the next morning when I awoke, and not long after, was turned on my back again. And this time my knees were operated on.

The knee operations were fascinating; as was the stomach one. And I could feel and sense all the devices or tools being used on the outside and inside of my body. It was pure wizardry. And I remember very clearly what was done on my knees. First, there was something pinned through the centre of my kneecaps, like they were being screwed down. Then followed stitching type work, being done around the outside of and under the kneecaps and surrounding area, and this was a different kind of stitching, with a different type of device – something different again.

I certainly couldn't see these tools or devices that were being used; to the human eye they were invisible. But I could feel and sense and quite clearly define their structure by what was being done. This gave me an almost clear picture of the type of device being used. Some of them were no different from the type of things we would be familiar with, only more sophisticated, or different versions of them; and which worked machine like and rhythmically, and did things to perfection. And there were other things being used with them that I couldn't describe, but were made up of sensations and rhythmic energy's. All of these things were interacting and working together; the whole process flowed, and had a magical quality to it.

Later when I got out of bed: and not sure what to do about my knees? 'Whether, I should still strap them up or not. And after thinking about it, I decided I would still strap them up as usual. And then not long after this, as with the previous day, the same pattern began to unfold; with me being put through physiotherapy and exercises on my upper body and legs.

And it was also later that same day that my eyes first started to be worked on. This started by having me first take off my glasses, and then being made to lie on the floor in the living room; on my back: and in a position where my head was held still with my own hands. And it was then that parts of the sun's rays coming through the window were formed into what seemed like a type of laser: which I could clearly see, and which was then used to do work on my eyes; and which lasted about half an hour. This was followed by eye exercises being done, my eyes being moved for me. This initially', while I lay on my back, on the living room floor, with my eyes being made to focus on a corner point, where the ceiling and walls meet. And from there, would be moved along the line that divides the ceiling and wall, to the corner point on the other side; back and forth from one side of the room to the other, and then up and down the corner line where the walls meet – from the floor to ceiling: many repetitions of them.

Once again I was fascinated by this latest operation and the method of it, and patiently lay there as this was carried out. And as for the eye exercises there were many more variations of them. And later that day when I went out they were put in motion. And they mainly focused on the use of the four corner points of large buildings; rotating in a circular motion from one corner to the other, while all the time keeping my eyes focused on the outside edge of the building – going to all four points either in a circular movement from the bottom corner point to the top one, then across and down to the other bottom corner and then back across to the beginning – doing them one way and then back in the opposite direction, or sometimes just across from side to side, or up or down: repeating these several times over. Other objects being put to use were lampposts; my eyes being focused on them and then being moved from top to bottom. And all of these would often be done at various distances, as it seemed that sometimes I was being stopped at a specific spot, sometimes close to the favoured object and other times at a much further distance from it. And what also surprised me about all this, was that they were similar variations of the eye exercises and techniques that the Bates Method taught.

But it wasn't just my eyes being seen to: when outside a further development of the physiotherapy on my body, was that this physiotherapy and exercises I was also being put through, now seemed to continue every day, wherever I was, and wherever it was convenient to do them; working on my whole body. Whether it was at home or outside, or waiting at a bus stop; it was being done wherever it was possible. And after just over a week,' I could move nearly all of my upper body normally again. The only part that didn't feel like it was completely healed was that same main bad area of my spine, and which was continuing to be worked on. And even towards the end of the second week this was still the case, this area of my spine and neck not yet finished with; this intricate work there was still being done.

This more intricate work on it still continued, and now sometimes lasting an hour at a time, with me being frozen in set positions while it was being done. These more longer types of operations occurring on about five occasions, and were excruciatingly painful. With my head often being moved and positioned and fixed in that position while this was being done. And often in-between these positioning's, rolling of the head and upper body and neck exercises were done.

I was by now starting to get a bit sick of these operations on my spine, and started to dread them; overwhelmingly because of the amount of pain. But also was aware and thought; that whatever it was that was helping me would eventually finish the work on that area, and it was a small price to pay.

I couldn't complain, and nor did I; Just a fortnight ago I was on deaths door. And now something had saved me and was putting me back together! In fact, what did strike me most about all the operations that were done on me, except for the ones on my spine, was that none of them were painful, but quite the opposite – painless!

It was also towards the end of the second week that the process to rebalance the hypersensitivity problem with my hearing had begun. This was initially started while I had been outside, i.e. walking past a shop where music was blaring out from. My initial and usual response was to put my fingers in my ears, but this time felt like I had been stopped from doing that and just casually walked by; and not knowing what to expect.

Only just under two weeks ago this too would have been unthinkable and also unbearable to my ears. But as I walked past, the healing energy began to swirl around me and furiously pump up through the palms of my hands and up into my head, and from what I later realized, was going into the hearing mechanism. And this would now happen whenever I was near any music playing through a speaker. This initial process first started in Oxford Street while walking past one of the big music stores, and continued and then progressed in other ways. The next step being: me going into places where music was on low or at a moderate level; like cafes or coffee shops, and staying for half an hour or so. It seemed like a step by step reprogramming of my hearing was being done, to rebalance and restore my ability to be able to tolerate all the different levels and frequencies of sound again.

On first going into a coffee shop and after ordering a cappuccino and sitting down, and not having to worry about the music that was playing; firstly: apart from the initial apprehension, it was something I hadn't been able to do for two and a half years, and seemed very strange. Just sitting there being able to listen to music again was strange enough. And secondly, it was a sheer joy and pleasure to be able to listen to music again. And what seemed ordinary and of no consequence to those around me, to me seemed so special, it was a real high. And not only because of the music, but the realisation' that this particular problem which had caused me so much hell, was now coming to an end.

It was also around this two week mark that other even stranger developments unfolded; namely: as well as the miraculous healings I was undergoing I'd also started to receive guidance in a variety of ways that seemed to compliment further what was already being done for me. As I had said earlier,' on the night when this had first started, while sat in the kitchen in my flat; and when later I had stood up and gone into the living room, and felt like I had been manoeuvred there, guided there; and by this second week, this had now developed into a situation where a thought would be projected into my mind in order for me to know when to do things. For example, the real sense of realising this was two days before the eye operations. "I was being directed to take off my glasses at certain intervals throughout the day and keep them off for varying lengths of time; and then similarly directed to put them back on. And before this: also during the long bouts of the loosening up and physiotherapy exercises, I was being directed to drink water at certain intervals by the same process; the thought appearing in my mind. This was intermingled with all the other stuff going on, the exercises and physiotherapy and operations etc, which I was now also experiencing similar directions for, and of the same kind; whether this was for me to take a break or go out for some reason, or go into somewhere, or something else! And as I soon realized was all for the benefit of getting me better. And by now this had become a regular form of guidance, with the thoughts clearly appearing bolder in a way that was noticeably distinguishable from my own; and with me being able to tell immediately. And as well as this, a further development of this was another form of communicating being produced! And this taking the form of a pulsating beat in the centre of my forehead.' It first occurring while in a health food shop I'd gone into to buy a sandwich; then while looking at the sandwiches on display, my eyes were took control of, and focused on and moved along the various sandwiches', in a scanning type procedure. Which comprised' of a circular motion of concentrated energy, rotating, simultaneously in front of both eyes. And which then my eyes were moved along them – as though scanning them all. Then locking my eyes on one of the sandwiches' a pulsating beat was then produced in the centre of my forehead, obviously indicating to me to buy that particular sandwich, which I bought. And as the days went by there was more of this particular type of guidance, guiding me to buy certain types of food.

Also that same day while I was in that health food shop there was also scanning over the various bottles of vitamins, and then the pulsating beat picking out some of them for me to buy too.

It now seemed like I was getting the full treatment. Not only was I being healed in a miraculous way, I was also being guided to buy the most nutritious food and vitamin supplements to compliment the whole process.

I remember the day following this new development, I had had an appointment to see my social worker, and wasn't sure if I should tell her everything that had been happening, or wait a while. And as it was such an unusual story and situation I didn't know what I would say or how it would come across, and especially with having the label schizophrenia on me, I didn't know what she would think either: whether I would be taken seriously was highly questionable. I called Margaret Palmer and asked her what she thought I should do? "She told me it was best not to say anything about it right now as they would probably think I had flipped. I thought it was the right advice; thinking, that telling this story and all that was happening I doubt I would have been taken seriously. And once I told them, I would have to stick to my story. And they would probably think it was a delusional faze anyway, or what they would term a relapse and may have me see the psychiatrist, or worse, in time, if I persisted with it, have me for a while admitted back onto the ward in the Psychiatric Unit. So for now, I decided to say nothing about it.

On seeing my social worker the next day, she had noticed how I had sat down, and remarked' that I seemed to have more movement in my back; noticing also that I hadn't used my walking stick to assist me when I'd sat down. I said yeah, "my back had become a lot better during the last two weeks. And remarking that the big improvement was due to me going swimming several times at the local swimming baths, and that this had made a big difference to the problem; which she replied that it was something of a miracle, or its quite miraculous - and me fully agreeing with her.

The truth about the swimming was that it had been suggested to me a while back by a physiotherapist who had said that swimming was good for back problems, and that I should give it a try, and out of desperation to try anything that may help it I tried it once; this some two months earlier and that was it, and at the time had told my social worker that I intended to go again.

That one session of swimming like other things I had tried, had slightly helped loosen up my back for a couple of hours, but then as usual it soon returned to its worse state. But I had been eager to give it a few more sessions and see if it could really make a difference. But as events unfolded the rapid deterioration of my back had stopped me from doing so: to the point where I was then having to use an extended device called the helping hand to put my socks on with, as well as do other things; picking things up etc.

And although I didn't like having to lie about something, I felt that for the time being it was necessary; for a number of reasons – the right thing to do.

It was some days later: now around the two and a half week mark' that the strapping for my knees was permanently dispensed with. This happening during the daytime in my flat, in my living room, after being put through a session of physiotherapy and exercises. It was a bit weird in how it was done, how the process or should I say the ritual, unfolded. Firstly, after I had been put through a session of physiotherapy and exercises; and still standing in the middle of the living room, I was then directed to go over to my walking stick which was stood up in the corner of the room; directed to pick it up, which I did; then went back and stood in the centre of the room. I was then directed to take off the strapping on my right knee. I peeled it off and threw it aside. The walking stick was then tapped three times against that knee; then soon after, brought up to my face and pushed against my lips; I suppose like kissing it. And then the same procedure was repeated for my left knee.' I was then turned around, and taken over to and made to look out of the window, and down in the direction of the rubbish bins; my eyes being made to focus right on them. The walking stick was raised up in front of me. I stared down at the bins, and automatically knew what this all meant. This was confirmation of my knees finally being cured and back to normal, my legs fully restored and no longer needing to use the strapping, and that I could now throw away the walking stick, bin it once and for all. However; it was a fold-up walking stick. And later, I chose instead to fold it up and put it in a shoe box along with the strapping and the specially moulded insoles, and the white noise generators; and keep them as souvenirs' from my voyage through hell. But at this moment, I stayed at the window; standing there, looking down at the rubbish bins, and, realizing what was taking place. Many thoughts came into my mind; mainly the many years of hell I'd endured with my legs. I stayed at the window for several more minutes; reflecting and pondering over the many thoughts before finally moving away. And although it was another thing to celebrate,' my thoughts and mood were more melancholy.

Only the day before, I had been back to see Margaret Palmer again. Things were now moving so fast, my health improving so rapidly, it was exciting. I was even thinking about the future. And as I was being healed of everything, I now knew that it would not be long, before I would no longer need to visit Margaret for the spiritual healing. It was on my last visit two weeks before, that I had made this appointment, and had had no idea of the extraordinary healing process that was about to unfold; and that during the following two weeks I would almost be cured of everything. I had not received any indication from whatever it was that was responsible for the miraculous healings and guidance, to not go back; or not continue with the spiritual healing that Margaret did on me. So I thought, until such a time or indication arises I will keep on going. And plus, I wanted to let her see for herself what had taken place, 'the transformation.' It was only two weeks earlier while there that I had sat in the same chair, still in a decrepit state and which I still had needed to use a walking stick to enable me to get back out of - virtually unable to move the top half of my body.

As I have said, the physiotherapy and exercises were now being done almost everywhere day and night wherever convenient. And it was this continued activity I was being put through day and night that had overwhelmingly been responsible for the rapid transformation. And it was not long after I arrived at Margaret's that not only did she see the transformation with my back and neck, but while sat in the chair having the spiritual healing this process of physiotherapy and exercises started up and she got to see it for herself. First with upper body and back exercises: then movements for my neck. Then shortly after I was manoeuvred up out of the chair and put through some more physiotherapy and exercises, for my legs and back, my body being taken over and all this being done for me right in front of her.

I think at this stage she wasn't as surprised as she might have been, as I had spoken to her three times on the phone during these two weeks and had kept her well informed about all of this. But now she could see it for herself.

After the session and display of physiotherapy and exercises, she had asked me more questions and even wondered or speculated that this whole thing may have come about as a result of all the spiritual healings I had had – or that she had done on me; and wondered if she may have even gradually grown more powerful as a healer,' as some healers do become more powerful, better channels, and that this may have come about as a result of that. Or that something in the spirit world may have been attracted to the healings and decided to help -- something like that maybe. And although I still didn't know what it was that was helping me, I couldn't rule out anything. Something was keeping an eye on me; 'and it had come from somewhere.'

It was a couple of days later; at the end of that third week' that I went to Wales for a break: now three weeks since this all began. And this whole process of help, healing and guidance I was getting continued. And now with some of the physiotherapy and exercises becoming more advanced, much faster and streamlined. And this included me now going for short runs: which would be put in motion sporadically throughout the day; and which usually lasted no more than a few minutes: two or three minutes or so; and my body was now starting to feel like a well oiled machine. I was feeling stronger and stronger, and it now seemed like I was as fit as I had always been: This in only three weeks. And also the feeling of freedom from no longer having to wear the strapping on my knees,' not having to tape them up every day, this felt great also.

I didn't really know how to explain to some members of my family what was happening, but had mentioned it to my sister two weeks earlier and mother a week before arriving in Wales, keeping it brief, but saying that something had been taking over my body' and healing me in a miraculous way. The previous few months, I had not mentioned and had stopped talking about how bad my health problems, and in particular my back had become; as I had become so depressed about it. And almost every time I had spoken to family members on the phone, it seemed like something had got worse, or a new health problem had developed. I'd got to the stage where I thought, what's the point' when asked how I was, in mentioning the latest miserable news. 'I got sick of hearing it myself! This had been going on for years. And had gradually cultivated the art of suffering in silence. Even the weeks leading up to me being brought to the brink of death, I had remained silent about how bad things had become; in the end just preferring it that way. Knowing that talking or moaning about it wasn't going to change anything. And so, while I gave some detail about what was happening I also kept it brief, wondering also what they would make of it all, if maybe they would question my sanity; or something like that. But what surprised me the most, was that they gave the impression of it being no big deal, and didn't really ask me in any detail about it; which I found a bit baffling, as what I was experiencing was so unique. And, it hadn't been that long since I last came for a visit and had quite visibly still relied on the use of a walking stick, to enable me to get in and out of chairs with, as well as do other things. And still wore glasses, and was now hardly wearing them. And couldn't sit in their living rooms if they had a television or radio on, and now it wasn't a problem – and now there was nothing wrong with me!

But I was philosophical about their response; and stayed focused on what really mattered. And concluded, that being healed of all these things,' was what was really important. And after a break there, and returning to London, there was now also some uncertainty about whether I would still be moving back to Wales. My family had assumed that I still was and up until this point I had planned to, and had some months back put these plans in motion, and which some family members had helped me with, to the point of finding a suitable house for me to stay in. But I was now in two minds whether to still go through with it. After all, my main motivation for moving back there had been the appalling situation I was in; and which I had now been rescued from. But even so,' there was still a part of me that still wanted to move back there. But it wasn't until a week after I came back to London that I finally decided once and for all, that I would not move, but stay put in London. As I still had ambitions I wanted to pursue, and this was the right place to pursue them in.

Although my eyes were still not yet fully back to normal, and were still being worked on in various ways: they had vastly improved, and now felt that they were about 60% right. And were still being attended to in the form of operation like things' still being done to them. These were not with lasers. But one of which comprised of something being used on the inside of the eyeballs - like a tiny rod, pushing away from the inside outwards. It seemed like it was changing the shape of the eyeballs; that's the impression I got, and how it felt. It gentle eased and pushed away at the inside of them. And although all this stuff was amazing and fascinating; I was now getting used to it, and nothing it did was surprising. It seemed that whatever it was that was helping me, and directing this healing energy or force, and movements; that when this energy was directed to heal something it would go into the body and do exactly what was needed' to heal that particular ailment or ailments, whether that comprised of the healing energy alone working on something or it turning itself into tools of some kind, or combining these with the movements and exercises. Whatever the problem was, it would do exactly what was needed to correct it, it was intelligent. And as well as the eye operations; the various eye exercises still continued to be done.

But although my eyes were still not yet fully back to normal, it had been only a week before I had gone to Wales that I had looked into the mirror again properly, for the first time in over five years, and had had a good look at myself. My eyes now able to remain focused for much longer periods. This was also very strange, and not just because of an absence of the usual symptoms; but being able to have a good look at myself again, and to see if I'd changed, or how I'd changed.

I didn't seem to have changed much, I still looked the same. And it was weird standing in front of a mirror again after all this time. And although I hadn't really changed, it was almost like looking at a stranger. And over the days following this, I often kept looking in the mirror again and again, almost fascinated that I could now look at myself; and by now was back to being able to shave, comb my hair, and do all the other things you would use a mirror for.

And also, while I had been in Wales I'd noticed that I could now look at the television again, the brightness of the screen no longer bothering my eyes. However, like I said, they were still not yet a 100% right and still not yet able to remain constantly focused on it in the normal way. But they were more robust, and I could now look at the screen longer than at any time since the eye problem began six years earlier. And this was without the glasses!

This was exciting, and I could now feel that my eye muscles were gaining strength by the day. And on the train journey to Wales I had bought some newspapers and when I had browsed through them noticed that I could stay focused on the small print and read some of it for about ten seconds at a time. And as well as the eye exercises that were still being done for me, I was soon made aware that my looking and reading through these newspapers, also was part of and an additional form of exercises for the eyes, to help strengthen the focusing – giving the eye muscles a workout. As I had found trying to focus on this size of print whether in a book a letter or newspaper, to read it had been almost impossible since the eye problem began, but this now too was being dealt with, and in this same way almost every day. And this being able to focus on small print again was very significant. And with this as well as the other eye exercises and ongoing work on my eyes, it now seemed that it wouldn't be to long before my eyes would also' be fully back to normal as well.

The progress was fast, and it was only three or four days later around the four week mark since this had all began, that I finally stopped having to wear the glasses once and for all. And although some exercises for my eyes still continued to be done, my sight and focusing now seemed normal again.

There was no ceremony in ridding myself of the glasses. I had been keeping them off for longer and longer periods of time. And on this particular day as I was about to go out, I went to pick the glasses up to take them with me - but instead, just decided to leave them. And from that day I never wore them again. My sight and focusing had now been improved to the point where I didn't need to. And as well as the things that I could now do again, because of the restoration of my eyes, not having to wear glasses anymore was also very liberating; and was another problem that now seemed cured.

I had felt this same sense of liberation' even more so, when for the first time in five years I didn't have to strap up my knees anymore. The real sense of liberation becoming clear the day after it was confirmed: when venturing outside for the first time in five years without the strapping on them. They felt bare, almost undressed, and as I walked about throughout the day I kept thinking that maybe they would go bad again, slide out of place, or, something else would happen; but it didn't; and by the end of the day I had realized that my legs and knees 'were' finally cured.

By now, I was becoming ever more curious' about what it was that was helping me. And a number of things had crossed my mind as to what it could be; from Angels, to God, or Spirit Guides of some sort. Anything was a possibility. And although in the past, I had been sceptical about the existence of some of these things; 'that was no longer the case. But, it was right around this four week mark; that I had a real contact of communication, and knew for sure what it was that was helping me.

This occurred late one evening during the continuing healing work still being done on my spine and neck; and on this occasion I was sat in the kitchen. The work that was still being done in this area, now seemed like it was coming to an end, with each session on my spine getting shorter, and to the point, that it now seemed that it wouldn't be long before my back and neck too was finally finished with and cured also.

Anyway, I was sat in the chair in the kitchen and had just been through what felt like finer, intricate work being done, in that same area of my spine. This was followed by rolling of my head and some neck exercises; and it seemed that this session had then finished for the night – or for the time being. And feeling relieved, I turned and pushed the back of my chair against the wall, and leaned right back into it.

I sat there in the kitchen, in the silence, and still of night; in thought, and thinking about the help I was receiving; and how amazing it all was, when suddenly, I felt my face being touched.

If this would have happened over a month ago I don't know how I would have reacted. But having and continuing to experience what I was already experiencing, although it seemed strange, it wasn't any stranger than what was already happening to me. Nevertheless, this was different, and it didn't stop there. I was first touched on my jaw close to my chin. Then gentle touches followed' along my jaw line; around to the side of my face, up to my forehead, across it, and down the other side: just like in the pattern of a horseshoe.

I wondered what was in the room with me. I certainly wasn't frightened, but felt comfortable; it almost seeming like a reassuring gesture; and I asked out loud, what are you or who are you, and waited, now expecting a response: But again, no reply.

There was obviously something in the room with me, and I looked up above me and around the kitchen, hoping that I might be able to see it; and at the same time, was also fascinated that something I couldn't see, could touch me and was in the room with me, and it could see me but I couldn't see it. And although I felt comfortable, there was also something slightly eerie about the situation too.

I stayed sat in the chair, and eventually relaxed, pondering over the situation, and wondering if this source was ever going to reveal itself. I suppose to some degree, it didn't matter if it never did reveal itself; as the most important thing was that I was being healed, cured of everything; and I suppose it wasn't that important. And as I pondered on in thought I had reached this conclusion; and my thoughts eventually drifted away from this; and into the peaceful quiet of night. Then when all of a sudden, a communication started: and the source of my good fortune revealed what it was. And this was done, with the now familiar, distinct, 'bold like' thoughts, flowing into my mind. And something else I hadn't yet experienced. This being: images being produced with them; being projected into my mind. And with a combination of these, slowly and carefully proceeded to tell me what had been helping me.

First telling me, that they were Angels, and that they were responsible for all the healings, operations and guidance – for everything!

On being told this and finally knowing, I didn't really know what to say; my response was a mixture of surprise and awe, and a flash of fascination, 'thinking' Angels do really exist. But another part of me wasn't surprised, thinking that what I had witnessed over the last month, I now thought that anything was possible, and thought also; that I had suspected that it could be something Angelic as well as other things. And, I was also told there were four of them, and they worked on shifts.

There wasn't anything more after that, it was just brief; they didn't tell me anything else. And when I did think of some additional things to ask, there was no response. I suppose they had told me what I wanted to know, and in a way that had further displayed their range of magical abilities. And it certainly made me feel special, to know that I had been saved and was being helped by Angels. And realized what I was experiencing was also something very unique. I was very lucky indeed. I felt blessed.

Chapter Six

The Trickery the Horror

And the Reality

Up to now everything had been going great, and I was almost back to normal. And the revelation of the previous evening of the Angelic presence had seemed like the icing on the cake. It was now mid-September, a little over four weeks since this had all began, and like I said everything was going great. The weather was still nice: warm and sunny. And on one sunny day around this time, I had decided to go to Richmond for the day; to have a look around the shops and have something to eat. And just laze about down by the river: which during the summer, when all is in bloom, is picturesque. And while there, I had browsed through some of the shops, 'mostly window shopping, and had come across a jewellery shop. 'A number of crosses on display had caught my attention. And as I had been looking at them, my eyes,' suddenly were taken control of; and manoeuvred over each one of them, in the same way that I was now used to -- the crosses were being scanned! I was thinking of buying one before this had occurred, but hadn't yet fully decided for sure. They were all silver crosses,' of different styles and sizes. The scanning went over all of them. Then eventually' locked my eyes on one. And then one single strong pulse was sent to the centre of my forehead.

I had now become accustomed to this pulsating beat, and it had gradually become just a single strong pulse' instead of the initial several continuous beats I had first experienced. I now knew what it meant and I think this was the reason why only one was now necessary. And although I hadn't fully made up my mind whether to buy a cross and chain, I now felt an obligation to buy it; thinking: maybe the Angels just wanted me to have it as a present or souvenir from them, something they had chosen, or picked out for me –

something of that nature. And I liked the look of it as well, so I went in the shop and bought it, and then put it on round my neck.

Up until now, I hadn't felt the need to question the Angels when they wanted me to buy or do certain things, or even go out of my flat. It all seemed like it was for my own benefit; and which it had been, and more than well proved to be so. So as well as liking the cross, I bought it without question; and it wasn't a cheap one either. I later had something to eat and spent a couple of hours by the river, enjoying the scenery and warm weather; before heading back to central London. But funny enough, the next day while at a street market at the Islington-Angel I ended up buying another cross and chain, but this one only costing me five pounds. I had been impressed and really liked the cross I had bought in Richmond, that over night I had become something of a fan of wearing a cross and chain; and seeing a selection of them at this street market stall, one had caught my eye, and as it was cheap, bought it on the spur of the moment: this I picked out on my own and nothing to do with the Angels.

It was later that same day, towards the late afternoon, after I had gone home; I was sitting at my kitchen table having something to eat. And was looking at the cross and chain that I had just bought. I'd put it on the kitchen table and was looking at it while I was eating. But, as my eyes had drifted away from it, they were suddenly took control of, then turned back and focused right on the cross and kept their focusing on it. And if I tried to look away, my eyes would automatically be moved back onto to it. This went on for about four minutes and by now I had finished what I was eating. It carried on and continued and, becoming increasingly curious, as to what was going on, 'I then started to feel a build up of energy in my stomach. Which quickly started rising up to my chest - and eventually making me be sick into the kitchen sink. It came up that fast, that I didn't even have time to run to the toilet: it would have gone all over the place, and had to quickly lean over the sink, which was right next to me. After I had been sick, I sat back down at the table, and was made to look at the cross again. The same thing happened, and I was sick again. 'I wondered what was going on. It seemed like the Angels were trying to get something out of me.'

I was then taken into the living room and made to look in a mirror on the wall' above the heater; which I had placed there a couple of days before. And as I looked in the mirror, my eyes were moved to the cross that was around my neck – the one I had bought in Richmond the day before. And as I stood there, staring in the mirror, my eyes fixed on this cross; gradually I felt this same slow build up of energy in my stomach again; and it started to rise up through my body. Then I felt a tremendous surge of energy rush up, and was thrown back about three feet.

I was then brought back to the mirror. And when I looked back into it; my face had changed into a nasty expression. I was then brought even closer to the mirror, and my eyes fixed back on the cross again. And as I looked at it, another tremendous surge of energy quickly rose up and I was thrown back again; but this time not as far. Once again I was brought close to the mirror, and my eyes were kept fixed on the cross, and were kept looking at it. And then to my 'horror', something started hissing through me, 'using my voice', hissing at the cross; and the expression on my face became nastier,' almost evil looking. And then I realized,' that there was something inside me, and that it was looking straight at me: and 'using my eyes to do so.

The hissing became more aggressive. It was frightening! And at the same time I was horrified, as it seemed like something straight out of a horror film. But I also felt some reassurance; knowing and thinking' that the Angels were obviously trying to bring it out of me. What it was, I don't know. But all kinds of thoughts ran through my mind and I just remember thinking that this is unreal I don't believe this is happening and that this is the kind of thing you see in films and are somewhat sceptical about. But it was real; and it was in me, looking right at me, and hissing at me with an evil expression; and to say I was horrified would be an understatement.

The scenario of me being made to look at this cross while standing in front of the mirror' went on: on and off all afternoon, the same pattern repeating itself. They were trying to bring it out of me,' but it wouldn't budge. And although I didn't want to believe or acknowledge it, something had obviously got inside of me, had invaded or possessed me, or whatever you want to call it. When or how or why, I don't know. But what also surprised me, was the use of the cross being put to use to drive it out. This was something else... which I had considered to be associated more with Dracula Films, and did not really feel that in reality it had any real effect on things of a demonic nature. But this, as I had just witnessed was not the case and from what I saw the cross was very powerful and did have an effect; and this itself was another revelation.

It was later that day, during the early evening that I phoned Margaret Palmer and told her what had happened, and she told me to come and see her the following day; saying that she could try some things that might shift it.

As well as knowledgeable in general of things of a spiritual nature, and from what I found out the next day when I went there, she was also knowledgeable about things of this nature too. And while I was there, she did some spiritual healing on me, as well as additional things to compliment it that would assist in the removal of this thing from my body. I was sat in the chair while this was going on; and as this was being performed' this energy started boiling away in me again, and then started to rise up through my body; and as it rose up further: up towards my neck' the hissing started again; followed by gurgling screaming type noises, and they got louder and louder; then finally stopped. At this point I felt that maybe it had come out and so did she and Margaret opened the window' so it could leave the room. But we soon realized it hadn't gone anywhere, it was still with me. This being confirmed a couple of minutes later by it making more noises when I checked by looking at the cross I'd had in my hand.

Margaret then told me about someone she knew who did something called 'Remote Spirit Release' and phoned her up for me right there and then. Her name was 'Janet Richards and as well as dealing with this kind of thing, was also a spiritual healer and a medium,' and apparently she could remove these things. I talked to her and told her the situation. She told me that I didn't need to come and see her in person, and that she could do this type of work from a distance; and after agreeing, then told me that first she had to dowse to find the best day to carry out this procedure. She did this and called me two days later, telling me the day that she would do it.

From what I later found out, this procedure of 'remote spirit release' involved the practitioner going into an altered state of awareness' and tuning into the person, and scanning them in order to detect any foreign presence. And then with a series of methods and the help of an assortment of spirit beings from the spirit realm, would proceed to instigate the removal of it. And this remote spirit release' was carried out two days later on Monday the 18th of September. But without success! She also sent me a report of what she had found. She said she had scanned my body and seen that it was black with an 'Entity' (a spirit of some kind) and that it had been with me for six years. She also said that she had scanned my flat and there had been some bad spirits in there as well, and that she had removed them. And after later confirming to her that it was still with me and that the remote release procedure had not been successful, she also tried again a couple of weeks later; and again without success. And by now this Entity had started to cause me problems -- it was trying to undermine the help I was getting from the Angels.

This Entity was very clever in what it could do. It started trying to mimic the healing methods, and the different types of methods that the Angels used' to communicate with me. But although it could mimic certain things; it didn't feel the same. I could tell the difference – or so I thought!

The healing methods that it tried to mimic were easily distinguishable as not being the same, not what I was used to. When I say healing methods, I mean the combination of physiotherapy and exercises and the other types of healing work still being done on me and was still being put through; although by now these were gradually tapering off towards a conclusion, becoming less and less, and it now seemed some of it was more a process of fine tuning. But this fine tuning process towards everything finally being a 100% right again was now being upset and hampered by this Entity. An example of this mimicking I first noticed when being put through some back and upper body exercises. As I had mentioned earlier, about how the tools and devices that operated on me, 'flowed, with a magical quality.' This was also true of the physiotherapy and exercises; they too had a special feel and flowing manner about them – as well as being precision perfect. And were more than just exercises'. There flowing manner was precise and machine like, and this is what I had become accustomed to. So when this Entity first started this mimicking, I immediately noticed something wasn't quite right; they felt substandard, almost clumsy, and without the usual special feel that accompanied them. But still surprising was the fact it could mimic these things – moving my body etc. And I said this to the Angels who after confirming it was the Entity doing this, said what it was actually trying to do, was stop me from getting better; and saying also that it had been the Entity which had been responsible for all of the health problems.

And as well as them telling me this; the Angels had also started to tell me a lot of other things by now; and the communication from them had became more open and they would answer most questions. But what they were now telling me regarding all the health problems was a revelation; and a shocking one at that. Also making me feel angry, thinking about all the damage this Entity had done to me and thinking,' no wonder nothing ever got better. I remember throughout the voyage through hell that it sometimes seemed that something was working against me, like something was stopping me from getting better. Now I knew the truth. I was right! Not only that, but it had been exactly six years when all these health problems started, and it was ironic that Janet Richards: although had not been able to get rid of it, had identified that it had joined me six years earlier: exactly when all these problems began.

It was also frightening to realize how clever this Entity was, and what it was capable of. This became more apparent by the day, when it sometimes succeeded in making my cured health problems a bit worse again; first doing this by trying to make my back tighter, trying to reverse the progress. My back was by now almost a 100% right, so I noticed straight away when this occurred. And what would follow, was the Angels would stop it doing what it was doing and put me right again -- reversing anything it had done, with the combination of the usual healing methods. It was also trying to do this to other things, my legs and eyes. And the Angels would have to do some work on them to put them right again also.

This same pattern continued. And it seemed like the Angels were also doing things to try and remove it from me. And, it started to become a battle between the Entity and the Angels. But although it was a strange situation, I remember at the time thinking that the Angels would eventually get rid of it. And felt somewhat reassured by knowing that; thinking, they obviously know what they're doing and have a plan of some sort to deal with it. I just had to be patient.

But time was moving on and the days going by: It was now the second week of October and nearly four weeks since I had been made aware of the presence of this Entity, and it still hadn't been removed. But the Angels, had obviously been monitoring its behaviour, and knew how it operated and what it thrived on. This was revealed to me one day while in my flat. The Angels telling me, that certain things in my flat were helping the Entity stay strong, by allowing it to draw energy from them.

It was on this same day that I was also for the first time introduced to a new method and further development in communication: The Angels producing a voice in my head – talking to me! It was bizarre to hear this voice in my head, and on first hearing it, it was shocking and surprised me. It was being blended in with the other methods of communication; and I was amazed by it, thinking, the things these Angels can do are incredible. It was a pleasant and very nice sounding voice, crystal clear and sounding like that of an upper-class English lady. And this combined with the other methods of communication that I was already used to, proceeded to tell me what these certain things were; 'and that I needed to remove them from my flat.

It was now the case that the Angels ways of communicating with me had become a sophisticated combination of methods.' Whether a clear image being projected into my mind's eye, or a thought or sentence being produced in my mind, like words being fed or inserted into it; or even my eyes just being moved and pointed at or shown something: And now, the addition of the voice in the head. I was guided and told things by all of these methods, often all of them working together; or whichever was the most suitable it seemed. If I asked something or needed to be told something or was being helped with something this is how it was done. And when this newest addition of the voice in the head appeared, it seemed like it was in response' to the emergency of the situation. And these certain things that I was told the Entity was drawing energy from', were "various objects of a certain colour," or colours. Being told as well, that some of the things I had brought into my flat were the influence of the Entity, influencing my thinking. And it was some of these things that also had some of these colours in them too. And that this Entity, was using, to draw energy from, and that I needed to get rid of them. These included objects of furniture and other items, from pottery I had made to pictures I had bought. And these colours that were the main problem: were anything that was red or black. The Angels telling me, that each and every colour gives off its own energy. And certain colours put together produce a powerful force for evil. And the combination of these two colours, 'red and black', together; was a very powerful force for evil; whether together or near each other. Telling me they had the same effect, and that together they were a dark force and attract dark forces; producing in my minds-eye', the image of the 'Nazi flag of The Third Reich' as just one example of this. And telling me the energy from these colours was fuelling its strength.

Not only was it enlightening to be told this about the colours, and surprising. But when I looked around the flat, there actually were quite a number of things that had these colours either in them, or were these colours. I was taken round the flat, and the things that had to be removed were all pointed out for me. First I was told to get rid of the red and black items: which I set about doing not long after. Even my double bed had to be removed; the mattress and box being a shade of red. Other items followed, and the process of stacking things up on the landing outside my flat door began, with my bed being put there first.

Later that same evening I was further informed that two other colours in the flat were also being used by the Entity to draw energy from. One of these other colours was a shade of green' and another was a light yellowish cream colour; and which one of the walls in my bedroom was painted in, right next to where my bed had been. The green colour was in quite a number of things. I was told some shades of green were o.k. but the brightest shades of green were the ones that were no good and had to be removed. My sofa had some green flowers on it, which were of this shade and therefore had to be removed, as did the underlay under the carpet which ran right through the whole flat; and was informed that the Entity was drawing energy from it even though it was under the carpet, and it had to be removed. So I ripped up the carpet, to get to the underlay and ended up just throwing out the lot; thinking that the carpet had passed its sell by date anyway.

But it was the next day, after having been out for a couple of hours and on arriving back at my flat that I was greeted by more paranormal phenomena. This during the evening of this second day of having spent two days going through my flat, purging it of all the items that the Angels had alerted me to. And towards that early evening had gone out for something to eat and had spent a couple of hours away from the flat. I needed to just get out and be outside for a while. I had had something to eat and went to a park for an hour and just relaxed in the quietness of the park and enjoyed the nice weather. And it was later, as darkness was creeping in that I left for home; then arrived back at my flat.

It was now dark. And it was as I opened my flat door and closed the door behind me, and then opened the living room door, then walked about four feet into the living room. 'It was then that a 'net' of some kind was thrown over me. 'And this was no ordinary net, made of ordinary material'. I clearly saw it as it was thrown, it was black, and when it hit me it crackled and cracked like a whip, then disappeared. It looked like it was made up of something of an electrical nature, like a bolt of lightning is. It was made up of something like that – seeming like an electrical net. And as it hit me and crackled, my instant reaction as well as being stunned and looking around - was what the fuck' was that!

This happened a few seconds after me switching on the living room light. It not only stunned me, but caught me by surprise. I quickly went into the kitchen - still nervously looking around. The Angels then told me what it was, saying: that there were other bad spirits in the flat, and because I now knew about them and there had been attempts to remove them, to get rid of them, that they were trying to frighten me out of the flat; and saying that they had been waiting for me.

It was scary stuff! And I thought,' that's all I need: "Anything else I should know about? I was also told that there were still more things that needed to be cleared out of the flat; being told that it was now an emergency to get the rest of this removed. Saying also that all the identified items and colours, clearing the flat of all this stuff, would also help get rid of these other spirits; they would leave – but not straight away. The next two days became a battle to get this done. And not only had the bad spirits tried to frighten me out of my flat, but the Entity was now also doing its best to stop me removing the rest of the stuff; and, I had to use the cross as a means of stopping it from being successful; keeping it at bay by sporadically having to focus on the cross for varying lengths of time, to ward it off so I could get this done: The Angels often reminding me to do this. By the fourth day I had got all of this done, and as I'd said, it was a battle. Now it was just a case of calling a rubbish removal service to come and take it all away.

This battle to remove the rest of these necessary items had gone on into the early hours of the morning. And some things I was told I needed to remove I had questioned; and not felt completely sure if I was doing the right thing, or if it was necessary. One of these things had been my screenplay, which I'd still harboured ambitions for and wanted to finish. So when the Angels told me that this' also had to be removed; telling me that a lot of it had been influenced by the Entity, and some of the dialogue and characters were evil. 'Saying it was laced with evil, and pointing out examples to me, 'I found this hard to believe. 'And was shocked, that I was being told that I had to destroy this too, to rip it up; saying to the Angels that I didn't believe it,' and that I could always alter it. But they insisted', that I needed to do this in order to be free of this thing: reminding me about the health problems and what it had done to me, adding also that if I wanted to be finally cured of Schizophrenia, it was necessary.

After a long time, and much debate, they finally convinced me it was the right thing to do, and in the end, I ripped it into pieces and put it in a rubbish bag and out onto the growing pile of items out on the landing. Another item or items of concern that I was told to get rid of were many photographs, stating that in some photos of me the Entity had changed my features or made me look a certain way.

These were shown to me and pointed out, and in the end I had to rip these up too, and felt equally unhappy about this also.

But anyway, now it was all done and my task on this fourth day was to phone some rubbish removal services to come and take it all away. I did this outside at a phone box; feeling relieved to get out of the flat. And after calling some services finally got one to come round the following day. Then the job would be completed, and that was that. And the following day as all the stuff was being loaded onto a wagon; after thinking about it I decided to save the bin bag which had my screenplay in, grabbing it off the pile and putting it in the bathroom on a shelf, out of sight; by now thinking I could alter it and that it was too precious to throw away; too much work had gone onto it. And when I had done this there was no complaint from the Angels. So I thought that maybe they will help me work something out with it.

Finally, when everything was removed and I had paid the people, I had a look around my flat. It was almost empty. Probably three quarters of the contents had been thrown out. And although I had for the last number of weeks been sleeping back in my bedroom, now I couldn't go back in there. Not only was there no longer a bed but the colour of the wall next to where the bed had been was a source of energy for the Entity and therefore I had to keep clear of it. I was now back to sleeping in the kitchen, but this time on the floor, on the sofa cushions I had kept and stripped of their coverings, and which were actually quite comfortable. But although I had undertaken the necessary steps the Angels had alerted me to and removed a source of strength for this Entity, also being told it would clear the flat of these other spirits; the danger was not yet over. And late in the afternoon on this same day was told by the Angels that I had to get out of the flat – for the time being. Telling me it was too dangerous to stay, and that I should leave as soon as possible; telling me to pack a small bag of items of clothing and toiletries etc, and get going.

They obviously knew something I didn't. I did what they said without question, and within twenty minutes I was gone. I've already said it was scary stuff, and even more so at this point. And although the Angels could do amazing things there were obviously other things out there that could do incredible things also, as had been demonstrated on me; and I suppose the seriousness of being told to get out of the flat was an indication of how dangerous these other spirits could be. And it now seemed that the most important thing right now had been to get out of the flat; as it had been stated that I was in danger. And the urgency of having to pack some necessities,' and the act of closing the door behind me and walking out into the street was met with huge relief.

Where I was going to go I didn't know! My years of the health problems had isolated me. I had no real friends as such' that I could ask to put me up for a few nights. Plus it was a very unusual situation, and I couldn't just go or be anywhere. It was now the case that I had to keep away from the certain colours that I had been alerted to. This all seemed part of a plan by the Angels to deprive the Entity of the energy from these colours; which would in time weaken it and which would enable it to be removed from me. I now had to avoid being near any of these colours, and anytime that I was the Angels would quickly alert me, showing me where they were so I could move away from them. So after leaving my flat, bag in hand, I walked down the road; not knowing where I would go. Another problem was, I didn't have a lot of money either.

I called Margaret Palmer to tell her the latest news, and told her what I'd had to do. She offered to borrow me some money, which I accepted and went round to pick it up two hours later. It was now dark. I was there only a minute and then left. And over the course, of the days following, I'd stayed the night at various places; first a cheap and grimy bed and breakfast in Earls Court, followed by two nights in an Indian YMCA. By the fourth night I did not have enough money left to afford another night in the YMCA, so that night I spent riding around on the all night buses; getting on ones that travelled the furthest so I could get some sleep. And during the days I would sit in coffee shops or churches or just walk the streets. By the fifth night I had tried to get a bed for the night in a homeless shelter that I'd found out about, but had not got there quick enough, they had already filled up their quota for the night – so it was another night on the buses; but at least they were warm. I was by now getting a bit sick of all this, and wondered how much longer I could keep this up for: as I also really needed a good night's sleep, only managing to scrape about four hours sleep for each of the last two nights. One good thing on the horizon was that my benefit payment was due the next day and that would give me some money; which would make the situation more bearable. However, I'd left my payment book in my flat and would have to go back there for it. And from what the Angels were now telling me,' it would be a very dangerous thing to have to do. But I had no choice really and in the end made the decision that that's what I was going to do, and with the Angels advising me on how to go about the task, the following day I set off back to my flat. And on the way there I had also decided while there that I would also grab the bin bag with the screenplay in. Thinking that with the time I had on my hands that I could start taping it back together.

On eventually arriving back at my road, and walking down it, and getting closer to the house; a nervousness took hold. I eventually came to the house, and braced myself for what I might encounter. I opened the gate, then the front door of the house, and slowly quietly went up the stairs, eventually coming to the top, until I was right outside my flat door. For six days I had walked the streets, unable to go back to my flat for fear of the unknown. But now I was outside my flat door, about to put the key in the door and go in. A sense of fear gripped me, the adrenalin started to pump. The Angels again went through what I had to do. The strategy was simple, 'to burst in quick. First into the bathroom, grab the bin bag, then rush into the kitchen, quickly get my payment book from the kitchen draw; then get out as fast as I could. I took a deep breath, and put the key in the door. I turned it, opening one lock fast, and then the other lock even faster. I flung the door open and into the bathroom I went, quickly grabbed the bin bag then rushed through the living room into the kitchen, and quickly opening the draw, fumbled through it looking for the payment book, and finally finding it rushed back through the living room. The air was thick with tension' and I could almost feel the presence of whatever it was that was in there. And although I had completed the task successfully and did so in a manner of lightning speed; the whole thing had felt like it had gone in slow motion, and finally exiting and locking the door behind me, and quickly going down the stairs and back into the street, a huge feeling of relief came over me. And as I walked away, I just thought to myself 'this is unbelievable.'

I later went to a park, and looked through the state that the screenplay was in; and realized, that it needed a lot of time and work, in order for it to be pieced back together again. It looked like it was in a thousand pieces. And I decided that under the current situation, it would be too much of a task to undertake. I had under estimated the amount of work that would be involved; and later called Margaret Palmer to ask if she would look after it for me, and later dropped it off at her house.

But at least I had some money now; and after three more days away, I was finally told by the Angels that I could now go back to my flat, it was now safe to do so. It was a relief to hear this and I couldn't get back there fast enough. And although it was good news to hear that those things had gone – the "Entity hadn't," it was still with me, and still causing problems. And, after a week of being back, there was still no change in the situation. And, several days after I had come back to my flat, the Angel communication', that came in the form of the voice in the head', which by now, had become a familiar form of communication: had started to become a bit bizarre.' And, had made me think more, about the nature' of these Angels.

This had initially started during the time that I'd had to stay away from the flat. And I had become a bit suspicious about them by the nature of some of the things that were said to me. However, after several days of being back in my flat the Angels had now informed me, that this Entity had been mimicking the voice in the head: producing the voice in the head also! And making me think' it was them. And said that they wouldn't be using the voice anymore, as it was hard for me to tell the difference, and anytime that I now heard it -- that it was the work of the Entity.

Some of the things that were said and had really aroused my suspicions were things of an explicit sexual nature, real hard core stuff. And although some of it was amusing; it didn't go hand in hand with what you would expect of Angels. And now that the Angels had informed me of the trickery, it seemed to make the Entity worse and the days following this revelation, the Entity had started using the voice in the head to terrorize me with. And by the manner and type of things it would say, it seemed like a right evil bastard. And as well as other things, it was telling me the spirits were still in the flat and they were going to get me. Saying every now and then and repeating, "were gonna get you". And the accent of the voice had changed, sounding more neutral and masculine.

My use of the cross was becoming more and more necessary, and at times like this and at others of a disruptive nature I was reminded by the Angels to focus on the cross, as this would make the Entity back off for a while. And during the times when away from my flat, I would have the cross and chain with me in my hand, the chain wrapped around the middle finger with the attached cross in the palm ready to ward off and put a stop to its annoyances, or games, or harmful abilities. I would often think, what a situation this is, it's unreal. And although I thought that the Angels would eventually get rid of this Entity, there were times I thought that it didn't seem like it was going to be any time soon; and decided I would also look for additional help. First going to see a Vicar at Saint James church in Piccadilly, several times; and which resulted in disappointment and no change to the situation. Then going to a place called the 'Collage of Psychic Studies' in South Kensington and making an appointment to see a person who dealt with these things, being told that she was very experienced. I remember the appointment well, it was a big disappointment. The lady in question I don't even remember her name, only that the ritual she was supposed to have carried out, had no effect. I again spoke to her over the phone a week later and she had almost forgotten who I was, and had to tell her the whole story again. She seemed to be on another planet. It was a total waste of time and money. But it had been later that same evening, on my way back from the Collage of Psychic Studies' after I had seen her, and on my way home as darkness fell, and while waiting at a bus stop. And after having a particular troublesome journey back, 'with the Entity causing problems. That I had been sat alone at this bus stop, with my eyes focused on the cross, in my palm, and a couple of people came and sat down next to me; a man and a woman who were together and who had noticed what I was doing. The man who was sat furthest away, then came over to me and asked me if I was alright? To which, I replied that I wasn't and was having to deal with an unusual situation. "I was having trouble with a bad spirit I told him, and elaborated on the situation, 'saying that I needed to keep focused on the cross, to stop it from damaging me: and saying also that it produces a voice in my head and says vicious and unpleasant things and tries to frighten me. And although it was a bizarre thing to tell someone: at this particular moment I didn't care about being so frank. And as far as I was concerned, I wasn't alright! And now that you've asked; as unusual as it is, this is the reason why. This was the frame of mind I was in.

But, by what he said next he surprised me. Saying that he too had the same problem with voices in his head and saying also that they used to do the same kinds of things to him, and next saying something that shocked me. He said, yeah' some of the things the voices would say used to frighten me, they would threaten me and say, "were gonna get you". And what was so eerie about it, was that he said it in exactly the same accent and nasty way that this Entity said it. And what was equally strange was that it had been saying that to me on and off the whole evening, "were gonna get you," and to meet someone at a bus stop and be told this was eerie. And when he said that, I said; that's exactly what it says to me as well -- in disbelief really.

He then went on to say that modern medicine had helped him and that I should get in touch with Mind' (the mental health organization) and saying that in the end he had realized and learnt that the voices in his head were just different parts of his personality, from all over the years he had grown up; and then elaborated on other aspects of having that problem. But what struck me most about him; was his eyes, they looked like pure evil, and I found it very hard to look him in the eye as he talked to me. Throughout the years I'd looked in to the eyes of a multitude and could always look anybody in the eye. But his eyes looked so evil; they didn't look normal. I'd never seen eyes like them. And I could only slightly glance in order to be polite and even that was difficult, as they not only looked pure evil but had an evil feel that also came from them. His personality and demeanour was pleasant and genuinely friendly, and as a good judge of character, I couldn't help but 'notice, that his eyes and personality didn't go together.

A bus eventually came and they had got on the bus with me and sat next to me, still talking to me and giving me advice until I got off at my stop. And I had concluded by what he had said, and additional things he had told me, that he himself was possessed; but didn't know it. But I didn't tell him that, he wouldn't have believed me anyway. After all, they probably thought I was mad and in need of help. And that was without me mentioning the Angels.

It was now the beginning of November and some good news was that I had received a letter informing me that the housing association was now ready to install the sound proofing in my flat. The workmen came round to measure up and then started installing it a week later and took them a week to complete; and the effect of it was a drastic. I could no longer hear the people in the flat below me. It felt like a concrete slab had been put in throughout the flat, shutting them out of my life. It was great not to be able to hear them anymore, or the Nebulizer machine which was barely audible now and not a problem. But apart from this the Entity had become more problematic; and during that first week of me coming back to my flat, I had also been told by the Angels to make large gold coloured crosses' and put one in every room in the flat. I did this, guided by them - cutting them from cardboard. The crosses measuring twelve inches long and covered in gold coloured wrapping foil.

Another thing I was told to do was buy a gold cross to put around my neck: being told it was more effective, more powerful than the silver one. I bought one, and tried it out on the Entity whenever it was causing problems. And sometimes I just used it on the Entity just for the hell of it; thinking about what it had done to me; bringing it close to my face, thinking: here you bastard have some of this. It didn't like the cross one bit, as the hissing that would sometimes follow would indicate. Although in reality the whole situation was often quite frightening; and other problems had started to surface. One day while in my flat and sitting in the kitchen I felt these rushes of energy rising from my feet and quickly going up my legs. The Angels told me what it was, saying that the Entity was trying to let more spirits into my body, and that's what that had been. It felt really creepy and horrible, like something entering the soles of your feet and making its way up, inside your legs. I was told to look at the cross, and from now on to always sit with my legs crossed at the ankles; to stop it from happening.

By what this Entity had done to me and was still trying to do I realized it was obviously evil. 'I also now had to use the cross almost every night in order for me to get to sleep,' and had to sleep with my legs crossed at the ankles; and even drew crosses on the soles of my feet. All in all it was right carry on, and like I said, often frightening.

I had still been seeing my social worker regularly and additionally to this was still attending appointments with a Psychologist at the Ear Nose and Throat Hospital. These appointments with the Psychologist' had started several months before, in May, three months before the magical whirlwind, and had been about every two weeks. Seeing the Psychologist had been the idea of the 'Hearing Therapist and Consultant.' In all reality because everything else they had tried on my ears had not worked, and I suspect the extreme nature of what I had described to them about my symptoms and with nothing really working on them, they had in the end probably thought it may partly be of a psychological nature too: Especially with me having the diagnosis of suffering from schizophrenia! Or it could just as well have been a way of passing me around. Whichever, it didn't matter, and as far my ears were concerned they were almost back to normal anyway. And the only reason I still kept up these appointments was because of the unpredictability of the situation I was in. And also because I had a lot of time on my hands,' spending it mostly on my own. And as the appointments were now only once every three weeks, or a month, I kept them up. It was just good to have somewhere to go to, and also have a talk with someone; it was a temporary diversion from the madness of the situation. And anyway, if I had told the Psychologist about the events of the last three months I doubt she would have believed me. And with me sitting in her office anxiously, sporadically staring at a cross in the palm of my hand, I'm sure she would have almost certainly put it down to mental illness. And even now when visiting my social worker' I would always have the cross in my hand, and while there would often have my eyes focused on it for long periods of time also.

There were many reasons why I would have to use the cross. One of them being to stop the Entity from drawing energy from the colours that gave it strength; the stronger it was the more damage it could do. So when I came into situations where I had no choice and had to stand, sit, or be near something of a particular colour, looking at the cross would stop it from drawing energy from them. This is what the Angels had instructed me to do; and was constantly reminded of its importance.

But it was during the middle of November, that out frustration, impatience, and often desperation that I had again contacted Janet Richards and spoken to her; telling her of the ongoing worsening situation! And that she had put me in touch with a man called 'Alan Sanderson. Who, as well as someone who also did remote spirit release,' also held courses teaching others how to carry out this procedure. She had told me that he was the most experienced person she knew. I contacted him immediately, and told him about the situation, and as he lived in London made an appointment to go and see him: And eventually having two appointments. The first on the 27th of November, the second a week later, and with the same outcome as before: he too was unable to get rid of it.

This was done under Hypnosis and he tape recorded the sessions. But although under hypnosis, I was fully aware of everything. He tried to find out what the Entity wanted, and asked for permission to allow the Entity to talk using my voice... which I agreed to. Initially the Entity answered some of his questions by nodding my head for yes, and shaking it for no. But obviously didn't like the questions and gradually descended into hissing, which continued on and off throughout the sessions. He wasn't put off by the hissing, he said he'd seen it before; and humoured the Entity by addressing it as Hisser'. The only word it did say was towards the end of the first session. When he said to it, that you thought you were never going to be found out didn't you; but you have been haven't you. The Entity then became a bit hysterical shouting yes yes yes yes, repeating it over and over.

I already knew it could talk through me, as only a few days after the Angels had first brought it to my attention,' and had me focused on the cross one night in my kitchen, while my eyes were focused on it, the Entity had started hysterically saying something, in what sounded like an oriental language and had me thinking at the time that it was the spirit of an oriental person. But as time went on and judging by the cleverness of this Entity I was made to realize it was just playing games, tricking me. And this Alan Sanderson had said also that these spirits were often tricksters; and funny enough, as well as working in this field he was also a 'Consultant Psychiatrist.

But although he had not been able to remove it, he put me in touch with someone in Canada who practised another form of remote spirit release. And after getting in touch with him and doing some necessary things including sending a photo of myself he too carried out his methods, but they too' were unsuccessful. And so my latest attempts to remove this thing from me had amounted to nothing.

Now December: And still no change! Only new more restricting' developments. Now the Angels telling me that I could no longer eat certain foods, or drink tea or coffee, as the Entity was drawing this same energy from them; and was now, very limited in what I could eat. I had also been told I could no longer go to a park either, because different types of greenery were sources of energy also. And although I had been adhering to all these things, and more, this Entity didn't appear to be getting any weaker, and, was doing its best to make my health worse. It would try to bring back the arthritis in my spine, and was also tying to produce it in other parts of my body, often producing it in my heals, which was extremely painful; and often took the Angels a long time, often hours' to put back to normal.

It seemed, like this Entity was constantly trying to undermine me; and I wondered how long I would have to live like this. It seemed to be dragging on and on. And on realizing more and more what this thing had done to me, I so much wanted the day to arrive when I would be free of it; and could embark on a new lease of life. I also wondered once that day came, what would happen to the Angels: Would they just disappear?

They had not said anything to me about this! Although they guided and cured me and communicated with me: they didn't answer all of my questions? And there seemed to be an air of mystery about their long term intentions. Sometimes I even worried that they would just leave.

But there were also further new developments in communication that I had also experienced around this time. I had been thinking about something, and had asked the Angels for an answer on what I should do. And instead of being told in one of the usual ways; they produced a gentle sounding yes by using my own voice to say it. It was very weird, and at first wondered if they had really done that or I had imagined it; but to confirm that it had been them, they did it again. This wasn't used often and only consisted of single words usually yes or no. No sentences, not yet anyway. Also I have to say I didn't particularly like this, it was to strange. And another form of communication that came around the same time, a bit later, and at first equally strange, was them moving my head, a slight gentle nodding for yes; and a slight motion from side to side for no.

I thought these new developments of communicating with me had in part come about to keep one step ahead of the Entity; as I had also been informed during this time, that some of the things that I had thrown out of my flat during the purge of items, had not been necessary. Some of which included photos tapes and CDS and an Indian cabinet. I had also been told; that the Entity had not only been able to mimic the voice in the head; but some of the other methods of communication as well. 'The Angels also now telling me' that not only did it have the power to mimic them, but to also block them out and make me believe that it' was them; and had done this during the later stage of the purge of all these items; and that it had taken them a long time to restore their link again. Pointing out to me that that had been the reason why they had let me rescue the screenplay, saying, the Entity had been responsible for making me rip it up. They went on to say that the photos, CDS, tapes and videos that had been thrown out had been the result of its mimicry also. This was especially depressing to hear. Purely for the simple reason that much of this stuff was irreplaceable - namely the many photos I had been tricked into ripping up, practically nearly every photograph I had, of not just me but with people I had known, and family members. I had been left with only seven photos. And especially depressing, was the loss of a video of myself when I was sixteen which contained a lot of footage of me in my first group. Also cassette tapes that contained music which I had played and sung on, playing songs I had wrote. Some of which was good stuff and which had been gathered over the years, some of it playing with other people and some on my own, and which I had thought about making an album out of. None of these things could be replaced. It was a great loss. To me it was priceless. I had at the time been tricked into thinking that much of it was tinged with the Entity's influence, and that I needed to remove all of it out of my life in order to eventually be free of it, being conned and bribed and tricked into thinking it was necessary: being told as these items were being ripped up or thrown out that the past was the past and that they weren't important and it had to be done. The Indian cabinet was also something I was really fond of, and had been tricked into throwing out: Being told it was a shade' of red! This cabinet had a hand painted design on it, and had come from India. I had seen other ones similar to it but this stood out; and had also regarded this as a bad loss, and at the time had tried my best to keep it. The Angels now told me that the actual colour of the cabinet was not 'a shade of red' but was called Mary-rosé, and was a colour associated with spiritual protection and was more common in India. And that it had been a source of spiritual protection for the flat.

I had actually bought the cabinet just a year earlier, and when it was delivered wrapped up, only after unwrapping it did I discover they had sent me the wrong one. It was the same cabinet but the wrong colour. But with the aggravation involved in having to take it back, I decided instead to keep it, and gradually realized it was much nicer than the one I had chosen. And after being told about the special quality of this colour,' I was also lead to believe that it was no accident that it had ended up in my possession.

The loss of these items had caused me so much sadness. And this even before I knew I'd been tricked into throwing them out. And the days and weeks after I had done so it had caused me to be very depressed at their loss; finding it hard to believe it was necessary.

Christmas came and went, and the cold winter months had arrived: Now January 2001. And with no change in circumstances I decided to call Alan Sanderson again. And who in-turn told me about a man in Luton called Ranjid; saying he was a healer and also dealt with bad spirits, and that his methods might be more effective and worth trying. I called him, and making an appointment for the following week went to see him; and over the weeks that followed, had three more appointments. I was told he worked with spirit guides. And as well as himself present when he carried out the work, there were two other people to help in this process. And, during the gaps in-between these appointments I had also been given visualization exercises to do, that were in some way supposed to help this. But as the weeks went by, and having gone there four times, it didn't seem like this thing was going to budge. And although he told me to keep coming, I wasn't convinced enough that this was going to work, and so moved on.

The Angels had not objected to me seeking help in other forms, and I had often wondered if maybe they might be trying to use some of these people as a channel to work through, in a similar way that healers channel the divine energy through themselves. And had wondered if that was what the Angels might be trying to do; and why they hadn't stopped me from seeking help elsewhere. Another reason for them not saying anything, I suspected, was that they didn't want the Entity to know all of their intentions. It was a way of outwitting it. And although the Entity had been trying its best to reverse all the cured health problems and was the reason for the slow progress of my ears being fully cured; it was in fact, in February of 2001, that the Angels finally told me that my ears were now finally' fully cured and back to normal.

The final hurdle had been tolerance to loud sounds and noise. A good example of this was that I still hadn't been able to tolerate the noise levels of being in a cinema. I could still only tolerate that level of noise for a couple of minutes before having to move away from it. But now that was no longer the case. And as my eyes had also been given the all clear two months earlier in December, I celebrated this by going to the cinema; for the first time in six years: seeing three films in a week. And although I was dealing with a difficult situation, it was still a great experience; and also a positive reminder of how far things had come; and that progress was still being made.

But time was moving on. It was now March 2001. And it had been nearly six months now that I had been dealing with this Entity. And was increasingly feeling frustrated that no one had been able to remove it from me. And this same pattern' of aggravation, and disruption to my life it was causing, continued.' And I was also feeling more and more frustrated with the Angels. Thinking about all the amazing things they had done to me; but why they couldn't remove it, get rid of it? I thought, surely, with the powers they had at their disposal there must be something more they could do to get rid of it. And by now, although feeling more and more frustrated, and somewhat despondent, I had given up trying to get outside help to remove it; feeling,' that the Angels "must" have some sort of plan! That would in time' eventually get rid of it. So I stopped trying, and weathered the ups and downs of the situation, just trying to remain patient, feeling somewhat assured and safe knowing that the Angels were looking out for me; and protecting me from its presence and ability to do harm.

I still visited Margaret Palmer for the spiritual healing, although now, nothing needed healing. But it was beneficial in other ways. And during these times I'd had little money and she didn't charge me. And she had also seen the whole scenario develop; and it was good to be able to talk to someone about it; as the unpredictability of the situation I was in, was often distressing, 'to say the least; as I was affected by it on a daily basis.

But the real horror of the situation; presented itself to me one afternoon, in my flat; during the last week of March. And the 'truth' of the situation I was in, was revealed to me. I was stood in the kitchen at the time, looking in a mirror on one of the walls; when I was shown and made to realize; that all the help I was getting and had got from the Angels: The Healings, the Communication and Guidance. All of these amazing things! And the Trickery and Mimicry and all of the things I had associated with the Entity; were all from the same thing -- the same source: namely, "the Entity".

When this was revealed to me, it initially filled me with horror; and then I thought that maybe I was just being tricked. But as the next couple of days unfolded, it was demonstrated to me time and time again that it was true. This was demonstrated to me in a variety of ways. And as a way of proof,' it would do all the usual things I was use to and associated with the Angels. Producing for me some of the methods of communication, as well as displays of the various forms of physiotherapy and healing. Then it would do all the things I associated with the Entity: the hissing' the not so convincing display' of mimicking the physiotherapy etc. And at the same time, making

126

me realize that there never were any Angels; and it, the Entity, had been playing a dual role, and had been tricking me all along. And it also confirmed to me, that it had been it that had performed all of the magical operations on me - and healed me of all the health problems.

During these days it pointed out and demonstrated many things that had been just pure trickery, including all the business with the crosses; making me look at them to see there was no real effect, 'to make sure I was convinced.' And all the stuff about the colours had also been trickery, just rubbish; and so many other things, the business with the bad spirits in the flat. It had just played on what Janet Richards had said about their being spirits in my flat. And it had also been responsible for the net that hit me; and the magical whirlwind that had saved me: producing the swirling type energy to further convince me of this reality. And also saying that it had been it', that had marched me up the hill too. This went on for four days. Until there was absolutely no doubt in my mind that there we're no Angels helping me; and never was. The Entity had been responsible for everything; "good and bad!"

It took several more days for it all to sink in. And I was in shock; and felt isolated. Thinking: what am I going to do now, now I've had it!

Whatever this thing was it was obviously very powerful; and had wonderful healing abilities, or access to them; as well as abilities to cause great harm; and seemed predominately evil. 'It was a real life horror story'. I also have to say that during the previous three months, I had at times become increasingly suspicious that the Angels and the Entity were the same thing, and had asked what I had then thought were Angels about this, saying, "that it sometimes seemed like you were both the same thing and seemed like you were inside of me rather than outside of me; but had been reassured by what I was lead to believe were Angels, that they were around me and outside of me.

But although some of the things that had made me suspect this were justified, I had no reason to doubt what I was told, as the range of abilities' of the things I thought were being done by Angels, were incredible. So on finally being enlightened to the truth, genuine shock took hold of me.

But this wasn't the only shock. I was also put in a terrifying situation in which this Entity told me that if I try to get help to remove it that it would kill me 'with a heart attack.' And it demonstrated its capacity to do this, by causing pains in my chest and heart, and its first demonstration was quite frightening. And then, whenever I would be thinking about ways to or what I could do to get rid of it, it would cause a sudden pain or pains in my chest area to remind me of this. This would be followed by a slight nod of the head, 'as a reminder; as if to say, 'watch it.'

This thing was always there, and it knew everything I was thinking; and this nod of the head, it wasn't just a nod of the head with no feeling – not just a movement of my head, but it would do this in a way that it fully imposed itself on me, so I could feel the personality of it coming through too. Like it took over my body for that brief time and function; being able to use it in the same way I could, so it was absolutely clear. This was eerie enough, but as I later learned this powerful Spirit, Entity, whatever it was, could do this to any part of my body; being able to move it in exactly the same way I could.

During the next two months, the realization that I was stuck with this thing caused me severe bouts of depression. It was a depressing situation in general. I did my best to avoid talking to it, and the only times I did, were due to frustration and anger at what it was doing -- and I didn't trust it either.

During that last week of March that it had revealed the truth to me, it was the week following that it had also told me that it' had also been the cause of the Schizophrenia too. And saying to me, that I no longer needed to take the medication that I was still taking for it; telling me the tablets were useless, and, that it,' had been the cause of all the symptoms that I thought and was told were mental illness. And that the only reason that I had thought the medication had made me better was because it had modified its behaviour, in order to make me and everyone concerned think that the medication was having an effect. Saying as well, that I had never been or never was mentally ill; and that it' had caused all the delusions, hallucinations, paranoia, and had been responsible for the evil thoughts and images that flashed into my mind; as well as some of the severe bouts of depression. Reminding me also' of a whole variety, of what seemed like real life' moving things or objects that I would often see out of the corner of my eye; but when turning my head to look at them "they did not exist." It also proved some of this by creating hallucinations right before my very eyes, creating several balls of light moving around on the walls, followed by changing one of the crosses on the wall into two crosses and then back into one; and creating an image of a face in the wall and piece of furniture. It was another demonstration of the incredible things it could do; and had done and said enough in the end to convince me that it was the truth; telling me also that it could turn all the symptoms and things on and off at will, and produce them at the drop of a hat.

I had been witness already to its ability's and the range of things it could do, and was capable of, and was left in no doubt that this was the truth; and from that day on stopped taking the medication. And there were no effects or side effects from stopping taking it, nothing! It had no effect on me. The Entity had also said to me that if it had wanted to, it could neutralize any effect of the tablets anyway.

On the plus side it actually felt good to not have to take the medication anymore, 'and know the truth about this. And one consolation was I now knew the truth about everything, and was also fully cured of all the health problems: which it had caused – and cured me of. But what next: What did it want?

During these couple of months after the full horror and truth had been established, they were to some degree quite a quiet time. There wasn't much communication from the Entity to me; except its little ways of threatening and 'reminding me of the consequences of me trying to get rid of it. And this was done with the other methods of communication, and not the voice. It had ceased to use the voice in the head. I hadn't heard it for quite a while. And plus I didn't want to talk to this Entity anyway; knowing what it had done to me, and put me through -- and was now' putting me through. At one point I had become so distressed that I had told my social worker about the Entity. Although, not the whole story about the healings and guidance – or what I had thought were Angels, or how everything had happened and unfolded; but just about the present situation' and the Entity threatening to kill me with a heart attack. My reasoning behind this, being, that this initially itself was enough to believe; and that if I started trying to tell the whole story, it would just be too much, and be perceived as being something else, delusional etc.

But it hadn't been my intention to tell her, not yet anyway. But had come about' during one of the weekly meetings with her, as a result of being so distressed and depressed, and it had just came out. I also mentioned it the next time I saw her; and was asked if I wanted to see the Psychiatrist and talk about it with her. I knew full well that the Psychiatrist could not help me, but went along with her suggestion, and agreed to talk to the Psychiatrist anyway, seeing her twice in two weeks. And elaborating and telling her more, and about the situation I was in, and about the threats and pains in my heart and chest that the Entity would cause. And on the second time I had seen her, she said that she didn't believe that I was possessed by an Entity,' and offered to change the medication I was on, (and which I was no longer taking anyway) to a newer and supposedly better drug called 'Quetiepine. Which although I knew was a waste of time and had no intention of taking it, just went along with it; just quietly thinking to myself and realizing what I was up against; and thinking that I should of just kept my mouth shut.

I had also written a letter for my family, telling them what was happening. Putting the letter in an envelope and on the kitchen wall, to tell the story and say goodbye, just in case the Entity carried out its threat.

Anyway, the Psychiatrist then said she would see me again in a few weeks; after giving the medication a chance to work, and see how I was then. And she also wanted me to arrange to see my doctor to have my heart examined.

But it was later, around this same time, about the first week in June, that the Entity had managed, over a period of a week, to gain my confidence, and tricked me into thinking the reason it was here, was because it wanted me to be a 'healer; and channel,' for the magical healing ability's it had – use them to heal people. I fell for it, thinking it really would be something to be able to show this to people, and heal them of anything. And after all, I had thought, it must want something; and this was probably it. And it was a better alternative to being threatened with death. It even helped me put an advertisement together ready to put in a newspaper, offering this amazing healing of any illness for a small fee, which would not have to be paid unless they were healed of whatever they were afflicted with.

It sounded fair enough! But my first port of call, and to try it out, and see if the Entity was being genuine: was to go to Wales.

My mother who had some months earlier fallen and broken her hip and was waiting for an operation for a hip replacement was an ideal candidate. So off to Wales I went, excited not only that I now had a career, but that I would also be able to heal my mother's hip, and do so to perfection, and that she would not need to have an operation. I called my mother beforehand, telling her that I was going to try this healing on her and that it could fix her hip.

But on the day that I did try it out on her, nothing happened. I tried two more times, doing what the Entity had told me to do, placing my hands near the area and waiting for something to happen. But, the only something that did eventually happen, was that I realized that this bastard, had cleverly conned me again: "Just another load of trickery."

Anyway, summer had soon arrived. It had been three months now since the truth of the situation was revealed to me. And as well as trying to do things to occupy my time – mainly as a distraction; I had also been buying some basic furniture for my flat; which had been left almost bare because of the Entity's trickery. But I was now also getting sick of this whole situation. It was now the beginning of July; and the frustration had been building up, to the point where one day I thought I can't live like this, this thing threatening to kill me, and playing games with my life; and thought rather than live like this I would rather die, 'at least go down fighting', and decided to do just that, and call its bluff and see what happened. Deciding to find out and seek whatever help was available in the field of the paranormal.

The first thing I did was go to the White Eagle Lodge; where I had gone for spiritual healing three years earlier, and see what help they could offer. They recommended the spiritual healing again, but this time using different colours; telling me that they were to cleanse the Aura (the energy field that emanates from and surrounds the body) telling me that the spirit was in the Aura. And so I started going there twice a week for this. But I didn't stop there and rely exclusively on this; I decided to contact other people who might also be able to help. One of them was a lady called Diana Cooper, a healer of some-sort, and who had written several books,' mainly about Angels.

About five years earlier, about a year after the health problems had started and my eyes were in a bad way, listening to the radio one night before the ear problem developed I had heard her being interviewed on a show: about how it was possible to bring Angels into your life to help you. And I had bought a cassette tape of her book 'A Little Light on Angels' which contained the necessary exercises to bring this about; hoping that they would be able to help my eyes. It seemed like a desperate and offbeat thing to have to do, but such was the hell I was going through with my eyes that I was trying anything. I did all the necessary exercises, several times, but nothing happened. But found the book interesting, and remembered how she had explained how she worked with Angels during the times that she had healed people of things, of both a physical and emotional nature, elaborating on how the Angels had taken things out of people. And now I thought that it was worth trying to contact her to see if she could help me. So I wrote a letter to her publisher to send on to her, explaining the situation; and within a week had received a letter back: dated July 26th 2001, and saying, that she was sorry that I had been invaded by an Entity, with such horrible consequences. But, also saying that she did not deal with them, and instead gave me the name and phone number of someone who does: but who lived in Scotland; and saying that he could do this work over the phone.

I was very disappointed that she couldn't do anything for me, and as I had already had people do this same type of 'remote spirit release' from a distance, and it hadn't worked, decided for now to put that on the back burner and see what else was available. No amount of money was too high to get rid of this thing. But I didn't have a lot of it, and as all these things cost money, what I had, I had to use wisely. And that meant trying something new.

Another person I had contacted and was prominent in this field was a man called Terry O Sullivan, who had written a book called

'Soul Rescuers'. Margaret Palmer had told me about him two months earlier, and had borrowed me the book. I had read it. It went into several areas to do with the paranormal' and spirits and possession etc, and how he had dealt with different things. From what I had gathered and been told, he was one of the top people in the world in this line of work, and was worth contacting. And as well as this, I was also told he was a psychic. And the only reason I hadn't yet contacted him was because of the threats that the Entity had been making. However, I had now contacted him, and spoken to him on the phone, and was impressed by his knowledge. And had made an appointment' to go and see him during the third week of August at an address in London.

During the two weeks leading up to this appointment; the Entity had reintroduced the voice in the head. Once again' using it as an additional way of communicating with me, or to aggravate me with. It had gradually reintroduced the voice by first making comments on things, or subjects I was thinking about. Comments were often sarcastic, or sometimes helpful. Prior to this, I hadn't heard the voice for several months. It had ceased to use it. But now during these last couple of weeks I don't know what it was up to, whether it was trying to work its way back into my favour, I don't know. But it had said to me during the week leading up to the appointment with Terry O Sullivan, that he would not be able to remove it from me: saying to me that I was throwing my money in the bin. I didn't take much notice of that, and thought it may just be trying to stop me from going to see him, trying to put me off.

On the day of the appointment, and after arriving there, and after discussing more about the situation and the Entity; he then went ahead with his methods of removing these spirits. This session of work lasted about forty five minutes, and included the use of a pendulum. And during which time I was lying on a massage table; and which by the end, he'd said he had taken four spirits out of me and that the Entity had gone.

As I got up from the table, I hoped he was right. He then directed me to go and sit in a chair, and sat in one himself, not directly in front of me but more over towards my left. He asked how I felt: checking myself, I asked the Entity if it was still there, and it said with the voice in the head' "I am still here." I then told Terry O Sullivan, saying – that I had just asked it if it was still here and it 'said it was still here. To which he replied, 'that they can sometimes, still linger in the mind for weeks after.

I knew for sure it hadn't gone anywhere; and that this wasn't a case of it lingering in the mind either, and asked it again 'double checking'- are you still here? To which,' it responded, with a slight nod of my head; accompanied, with the usual feel of its personality that came through with it. And I told him what it had just done, and that it had not gone; and no sooner had I said that then the Entity crossed my eyes. And I said look at my eyes, what is it doing to my eyes? He said their crossed. And then they crossed in even further and then even further in, so I couldn't see; and then my face was turned towards him. I realized then what the Entity was doing. And said, it's showing you that it's still here. (And as I thought later, 'blatantly' showing him)! My eyes were then soon after returned back to their normal position. I could see he was taken aback, if not slightly unsettled, but more importantly now convinced that it was still with me. He then went on to say that it was an unusual situation, as this Entity had made me physically ill for many years, bringing me close to death and then turning around and dramatically healing me of everything. He said it was obviously very powerful, and didn't think it was good or bad; and thought that it was not working of its own accord; 'and that it has its masters in the spirit world. And also saying, that he thought it was under contract to stay with me for a specific purpose, and until that had been achieved, it wouldn't leave.

What he had said was interesting stuff, and made some sense. I had increasingly wondered why it had done what it had done, making me ill, incapacitating me for so many years, bringing me close to death, and then curing me; and then eventually revealing itself. Surely there must be a reason for this. But at the same time I had also thought the opposite, that maybe it was just evil and delighted in its own power. And although I left there very disappointed that he had not been able to remove it, I had been impressed by the many things he had said. And thought that he had been the most knowledgeable person I had come across in this field.

It was later during the journey back home while on the bus that the Entity had reminded me of what it had said earlier; and saying to me: The only thing Terry O Sullivan had took from you was 75 quid -- and proceeded to flash images into my minds-eye of several much needed items that I could of bought with that money: a new pair of shoes, a new jacket etc, and finally flashing the moving image of the money being thrown into a rubbish bin.

Another thing I had done was contact Alan Sanderson again: and spoke about the new developments and ongoing situation; and he spoke to someone on my behalf called Lesley Moul, who I talked to on the phone and who told me about a breathing exercise I could do to strengthen the Aura and in doing so, supposedly would push the Entity out and away. So this was something else I had been trying. I had also written a letter and sent a photograph with it to someone in India called Sai Baba, who Margaret had told me about, saying he is a modern day miracle man and can do healings and various things from a distance. I had got the address from a book that I had recently bought, that was written about him and was also reading; and who I will talk more about later. And around this time I had also started to keep a diary of my life and this unusual situation. I was doing all I possibly could do; trying any source of help that was available. Another person from India that Margaret had told me about was a woman called Mother Amma; who was also a healer and revered spiritual teacher of some sort, and during September she had come to London for a week and was holding court at a sports centre in Crystal Palace, and I had gone there to see her. There was literally hundreds of people there, if not several thousand, and as well as other activities and side stalls the main highlights were her addressing the audience and speaking at length on issues of a spiritual nature. Later people lined up to be blessed by her and I had lined up with them and got blessed also. Like I said, I was trying anything.

I had been surprised that the Entity had let me contact, or see the various people I'd seen in order to remove it, and had not carried out its threat like it said it would; in fact, around this time it hadn't been too bad, it seemed to have become less of a problem. But nevertheless still a problem. And it was making other things more clear, telling me around this time that it had been with me much longer than six years, saying it had been with me from when I was a small child, even a baby. I found this hard to accept at first and didn't believe it. But its way of proving it to me over the days following this latest revelation was to remind me of many events in my life as far back as I could remember.

I initially thought when it was producing the events that maybe because of the abilities it has that it might just be going through my memory bank. But another way of it further convincing me was that it would make me have little accidents, like make me bang into something or someone, or step on someone's foot, or accidentally cut myself while using a knife, or even drop something; and it was amazing how fast and precise that it could make these things happen; and almost every time it would cause these things to happen, it would sarcastically say after, 'sorry about that'.

I thought at first that it was just another way of it wanting to annoy me or be nasty. But what it was cleverly doing, was making me aware of its ability to do this in order to further confirm that it had been with me as long as it said it had: reminding me after about four days of this, about lots of little accidents I had had throughout the years; saying that it had caused them, it had been responsible for them. And some of which had resulted in some very nasty cuts or other injuries. And something else that had afflicted me from about the age 13 until about 16, had been trouble with my sinuses, causing them to feel like they were blocked; and which I had gone to the Doctors for, several times, and eventually had to see a Consultant at a Hospital. This had caused me a lot of discomfort and lasted around three years, then just gradually went away and never troubled me again. But it was during these same four days, that another way of the Entity convincing me was to reproduce these same symptoms that I had suffered from all those years ago. And when this first happened my immediate reaction was that I recognized it straight away, 'and thought, I haven't had that problem for years. And as I was thinking about it, the Entity started repeatedly nodding my head, until it dawned on me that it' had just reproduced those same symptoms; and that it had been responsible also for original problem; and was showing me the proof.

All of this was produced as further evidence; until, like the last major revelation, it had finally convinced me that it was telling the truth. But that wasn't the end of it. It had also pointed out how it had been affecting me in other ways; subtly manipulating my thoughts, causing trouble in relationships, from, imaginary wrongdoings, to paranoia, to influencing decisions. It went on and on; and as well as being shocked, I gradually realized also that this thing had disrupted my whole life; and had caused all kinds of trouble throughout the years; and had influenced some important decisions in my life; which it took no time in elaborating on.

For me, it wasn't easy having to come to terms with this, often thinking from then on, how my life would have been different, and where would I be now minus its influence. And what would I have become. And another aspect of it was that it knew everything about me; likes, dislikes, thoughts, secrets etc, everything!

Another form of aggravation that had developed during this time, had been the use of its ability to take over and move any part of my body. It had been doing this 'sporadically' for a while; but recently had been doing it all the time. Examples of this are: when I was walking it would briefly take over the walking process so it would feel like it's not me who's doing the walking. Or when I would be making a cup of tea, it briefly would take over by picking up the kettle or stirring the tea. These were just some examples. But another, and more distressing of its abilities, which it had now made me aware of too' was its ability to change the way my face looked. This ranged from slight adjustments to its shape, to making me look much older, or my face look fat and bloated, or making it look weird, or unattractive, and even nasty looking. And it could change its expression' or way my face looked in a matter of seconds: and would demonstrate this to me when I was looking in a mirror.

I had over the years when sometimes seeing photos that had been taken of me wondered why sometimes they did not look like me, or the face I was use to looking at, and could never figure it out. And it wasn't just a case of a bad photo – as some of them looked weird. They didn't look like me at all. But this was another thing that had baffled me and I now had the answer to; the Entity revealing through its demonstrations that it had been the cause of this too.

Chapter Seven

The Revelations

My attempts to remove this thing from me had so far failed! I was living with a highly unusual situation: and would sometimes wake up in the morning and think, is this real? And to my horror the Entity would nod or wink or remind me in some way, that it was. I wondered how much longer I could go on like this, living with this situation. I felt like my life was not my own, and that my body had been hijacked; and there was nothing I could do about it. And I was increasingly feeling' that all of the people or so called experts who I had seen, and others that I had not yet seen did not have any real power to call on,' or the necessary knowledge to get rid of it; and felt like I was stuck with it. The feeling I got, was that this Entity was cleverer than them and knew things that they didn't know. And it had said after my latest attempt at freeing myself from it, that the only way I can get rid of it, is if I die, or it leaves me of its own free will: That really cheered me up.

Now the middle of September 2001; and six months since it had started revealing its horror story to me. And two months since I had – through frustration' decided to call its bluff and embark on a mission to free myself from its clutches. And the only thing that I was still doing and hoping would eventually have this effect, was continuing attending the White Eagle Lodge for the spiritual healing that they recommended; and usually going there twice a week. It was beneficial in other ways too, these being: having people there who I could talk to about it; as, it wasn't something you could talk to anyone about. Also during the middle of September I had gone to Wales for a week. For a much needed break from the insanity of my situation. And had told my mother and some other family members about what I was having to live with. And as I had explained to them how the whole situation had evolved, into the present horror story; it was now also becoming a longer' and a more complex story. And not wanting to burden them, I tried to play some of the worst aspects of

138

the horrors down: not saying it had 'threatened to kill me! And also tried to articulate what I did say in a way that didn't make me sound like a lunatic; but it was hard. And I had only really told them, out of being depressed about the whole thing. And not only that, I was also getting sick of having to tell this same story over and over again; and, the story was growing longer and longer. Not that I was just telling anybody, 'I mean the many people I had come into contact with in my efforts to get rid of it. And besides, there was nothing my family could do for me; it was something I had to live with and deal with by myself. I was on my own;' and fully aware of this harsh reality.

But as bizarre as the situation was! There were also times when this Entity would leave me alone, and not bother me; and this I had noticed was mostly when I was around people. So when this occurred; during these times I almost forgot it was there. And at other times, when with people' and I had forgotten about its presence it would do little things, or say something to remind me again, which I would just ignore or try to ignore. One thing it started to do which was aggravating was it would jerk my head 'turning it very fast towards my right shoulder, which to the observer would seem like a nervous twitch or tick; something my mother had commented on, asking me why I was doing that. And as well as this it would also cause my eyes to twitch or roll.

I had had a good break in Wales; and the Entity had not caused me too much trouble. And on returning to London, a week later, I had decided to call the person that Diana Cooper had recommended I contact. His name: Graham Wyllie, and who lived in Scotland; and who too, does what is known as 'remote spirit release. And while speaking to him was told, that he could do this work over the phone; but saying that he wouldn't tell me how he does it, as he didn't want the Entity to know. And saying that first I needed to write a detailed letter to him, telling him about this Entity: explaining how the whole situation evolved – the whole story. And the letter, which I wrote, although condensed, and cutting a long story short was some twenty odd pages long, and took me over a week to write; and after which I then sent him.

I received a phone call a week later, and arranged two half hour phone sessions; hoping that it may have an effect. But, the resulting two sessions, did not result in the Entity being removed.'

He too was knowledgeable, and knew a great deal about possession; and had said to me something that I was more than fully aware of; and that being: that some of these spirits' can play the body like a piano! And hearing him say this, had thought and said; 'that's exactly what it can do, "that's spot on." He further advised me to buy and read a book called 'The Power of Now' by Eghart Toll, and after I'd read it to call him back.

I did look this book up' and browsed through it, reading bits of it while in the book shop, but didn't feel it had much relevance to my own situation, and therefore didn't buy it or call him back again.

And again, before I had embarked on this latest attempt to remove this Entity, before I had had the phone sessions, the Entity had guaranteed a 100% that he would not be able to remove it; and was more or less mocking the whole thing, saying that's another £30 in the rubbish bin, and I'm wasting my money.

It had been around this same time,' that I had been thinking at what a unique and unusual story that this all was. And had thought that at some point,' I would like to write a book about it: And had even thought of a name for it. I was going to call it "Possessed," which I thought summed up the situation. But to only a week later change my mind. This because the Entity came up with a name that fitted like a glove – the name: The Unwanted Companion!

Literally this name did some it all up. It literally was' the unwanted companion – that was a fact. And not only was it a good name, it was spot on, and it was perfect. I thought yeah, that's exactly what you are, "the unwanted companion."

It was safe to say that during this time the Entity had become less trouble, and often' more helpful. This included, reminding me to do things, to even helping me write the long letter to Graham Wyllie: by it simplifying some of the things that were hard to explain. An example of this was that when I was writing, it would suddenly stop my hand, freeze it so I couldn't write, then tell me a better way of explaining what I was writing: making a point more specific; or adding more clarity to that sentence or paragraph, which I would then write.

It seemed strange that it had helped me with this letter, as its intended purpose was to help and assist in its removal from me. But it was through these, as well other demonstrations, that I had come to realize that this Entity didn't just have amazing abilities to heal or do other things, but was also highly intelligent and creative.

I had first become aware of this nine months earlier, back in January; when I had first started to sellotape the ripped up screenplay back together, and had noticed some of it was missing. And one scene in particular that had completely disappeared I'd decided to rewrite it; seeing if I could remember enough of it. And at the time, what I had then thought were Angels had helped me with this. It was four pages long, and it was better than the original. And I had been really pleased with it. But because of what the situation later evolved into, that was the last time I had worked on it'.

And as well as the letter it had helped me write, it had also started doing a similar thing when I had been writing in my diary; almost like helping me write it. It was a strange situation. And I had tried to refuse its help. But because its input was so good, and memory's of specific details. It enabled me to remember and write things much better – something which I had noticed! And often, 'though somewhat reluctantly' accepted its help for these reasons.

And now, as the Entity was seeming to become less of a problem, and appearing more helpful; even saying that I could benefit from its presence; and that, if I started writing the screenplay again,' that it would even help me with it? And it had been for these reasons also that I had accepted its help. Only a week earlier after it saying this, it had also said it could help me with the book I was thinking of writing, more or less telling me to get on with it; 'which had prompted me to think of a name for it. And only a week later for it to come up with the name, the unwanted companion.

But although it seemed like something had changed and the Entity was being sincere, I was still wary of believing it, as I knew' and had experienced its capacity for deceit and trickery before. But nevertheless, over the following days, decided to give it the benefit of the doubt' and attempt to start writing the book; and sitting down one evening to write the first page. Which I received no help for? But wasn't put off: as I hadn't expected it to help me straight away anyway.

It Was now the last week of October, beginning of November; and two nights after I had sat down to begin writing the book. I was feeling quite upbeat,' and had decided to go and see a Film in the West End.

I had arrived there in the late afternoon; and then looked at various cinemas to see what was on. And choosing a Film in one of them bought my ticket and then went into the auditorium. The seating area itself was half empty,' and the Film wasn't that good. But, I was just pleased to be out; and doing something that took my mind off the unusual situation I was in.

But it was while sat there, and half way through the Film, that the Entity surprised me by turning very nasty again. Using the voice in the head and with a very serious tone, it started saying it wasn't here to help me write a book, and it wasn't 'here' for my benefit; then carried on by saying abusive, horrible, derogatory and evil things about me: Then finally saying to me that this was my last night on earth. Saying, it had decided to kill me,' and that when I went to bed that night' that I wouldn't wake up again: it would snuff me out in my sleep!

'It had really convinced me it was going to kill me. I didn't know what to do. The abuse continued to the point that I left before the Film ended; and just walked out of the cinema feeling numb and drained. It wasn't like I could just call the police or something either. There was absolutely no one', who could help me. I literally was on my own.

I wandered the streets for a couple of hours, and sat in a fast food restaurant. Then around midnight set off back to my flat, the abuse and threats continuing on the journey home. And after arriving back at the flat, the abuse and threats now reached another level; with the Entity picking up the first sheet of paper I'd written for the book two nights earlier and ripping it up into small pieces, and saying while it did, that I'm not here for your benefit I'm not here to write a book, I'm here to kill you, and I've been waiting for this day for years and this is what I'm here to do. And it was saying this through me, using my voice, and at the same time looking at me in the kitchen mirror appearing angry at me and menacing.

It had taken over my whole body to do this; using in the same way I could. And I could do nothing about it. It was like two people, two personalities in one body, and both of them having the power to use it. I had not yet seen it do this before in this capacity, or this way, and this new demonstration only seemed to further confirm that it did fully intend to kill me; and that I wouldn't wake up again once I went to sleep. This went on for about half an hour after I had got home; then stopped; leaving me to ponder over my situation.

It was now around 2 a.m. and I was left to decide how I was going to deal with this. It was a terrifying position to be put in, and felt that I just needed to talk to someone, and needed to stay awake. I then got the number for the Samaritans and called them, spending an hour or more on the phone telling the story and the terrible situation I was now in. And after speaking to the first person, I then called again 15 minutes later, and then spoke to someone else for what seemed like a couple of hours.

It was now around 5 a.m. I had tried to stay awake, but in the end became so tired I couldn't stay awake any longer; and just went to bed -- and not knowing if I would wake up again. And when I did wake up later around mid-day; I was surprised I was still alive, and couldn't believe it. Thinking to myself, 'I'm still alive.' And the fact that it was day time; with the daylight' and the glow of sunlight streaming through the window; this had seemed to make it all feel that much better. And after getting up, and from the time I got up, the Entity had said nothing? And I certainly didn't invite it to say anything either; even trying to avoid thinking about it. But this absence didn't last long, only about three hours, by which time it had resumed its usual reminders of its presence; and later saying to me that it had postponed the day that it would kill me: saying it would kill me tonight instead.

This all happened on a Monday. And now it was saying it was going to carry out its threat the following night, which it obviously didn't: But was equally as scary as the night before. But it was during the Wednesday, that it said it would finally do it on Friday; and that that would be my last night on earth.

There was absolutely no doubt that it could snuff me out in my sleep; or cause me to have a heart attack, or death by a whole other number of methods. I had been made fully aware of its ability to do this, and it knew this, and used it as a way of terrorizing me. One thing, I had discovered' was being around people made me feel safer. And I decided to seek some refuge during the evenings of the following days, in a place in Islington called the 'Haven.' Which is a night time drop in facility open all night from 7 p.m. to 6 a.m. for people with mental health problems; who may be experiencing a difficult night or phase, and need somewhere to go during the evening hours, for either a chat with a counsellor; or to just sit in there in a safe environment with other people around. And it was during the Wednesday evening that I decided to go there. And with just being there I felt better, and safer. There was a common room where most people would sit and where there would also be a member of staff present. And if you wanted to have a private talk there was also a member of staff available for this. And was asked, not long after I arrived if I wanted to have a private talk with one of the counsellors, to which I said I did and later had a chat with someone.

I stayed there until the early hours, about 4 a.m. and did the same the following night. And by Friday, the Entity was telling me that this was finally the night; and even taunting me by singing the song by Rod Stewart, "Tonight's the night", sporadically repeating it in my head throughout the day, and emphasizing the lyrics 'tonight's the night'. And by Friday night not knowing what would happen, I again went to the Haven and stayed all night, and leaving around 4 a.m. again; and after getting home I again went to bed, and again: not knowing if I would wake up again. And what made it all the more eerie is that the Entity had not bothered me or said anything to me for most of the evening, and had remained silent for the past five hours, not even on the way home did it say anything, nothing! Almost like it had gone, or, it was remaining detached; ready to do me in my sleep. And again when later that day I awoke I was again surprised and also very relieved. But later in the day the Entity said to me that it had 'decided' to postpone killing me, for another week; and that it would be next Friday instead! To which I sarcastically said, 'oh thanks for that, that's very kind of you.

Sometimes I did get angry at it, and was sarcastic; and even abusive back to it; usually out of anger or frustration or other emotions it brought up in me as a result of its aggravations. But in general I had tried not to allow it to get to me or wined me up. And, as I had said about the name, 'the unwanted companion: it fitted like a glove, and it knew it.

Anyway, I had continued going to the Haven every night for the rest of that week, and noticing also that it was not the same staff on every night. And although I had been in an anxious and what could be described as a desperate state; when speaking to the counsellors, had not told them I was possessed by an evil spirit' and that it was threatening to kill me; as I knew only too well they would have almost certainly put this down as mental illness. And, even with the slim chance that they had believed me, what could they do? So I had instead told them that voices in my head had been threatening to kill me, and were terrifying me, and that I thought it was an evil spirit and was afraid to go to sleep; and didn't feel safe. Which in fact' was a large part of the truth; and, allowed me to at least talk about it and express how I was feeling; and get some of the anxiety off my chest; and also receive some help and advice.

Of that first week I had gone to the haven,' I hadn't felt the need to speak to a counsellor every night. But out of those six nights I had gone there, I had spoken to one on four of them. And it was on one of these nights, the fourth one, that I had had a good talk with one counsellor, who had told me that there was a lot of people experiencing the same problem as me, with voices in the head, threatening or saying nasty things etc, or telling them to do things. And further telling me that there was a 'hearing voices support group,' that is run by Mind (the mental health organization). And that it would be beneficial for me to go to it, and gave me the phone number.

I was glad to hear this; and was very curious too! And at the same time a part of me knew, that if they were experiencing the same or similar things as me: then they also, were possessed too; 'but probably didn't know it.'

"But it was something else she had said, that had really struck a chord. She said some of the people she had spoken to with this same problem had also said that they sometimes smelt strange smells or things no one else could smell. This was something I too had experienced over the years, and now knew that it was the Entity who had been doing this; and had responsible for producing them: another one of its many abilities which it had recently enlightened me to. And it's way of confirming and proving this, was to produce the same types of smells in different locations, at different times, and several times over; until it clicked and I got the message, that it had been the cause of these too. One smell in particular it had reproduced and reminded me of was the smell of gas! Some years earlier I had what I thought when one day entering the front door of the house and began to climb the stairs to my flat, thought I could smell gas. And could smell it in my flat also. I checked all the appliances and they were all switched off. So I went and asked the neighbours in the flats below me, seeing if it was coming from their flats; and could smell it as I was talking to them. And when they said they couldn't smell anything' was bemused, saying to them your joking aren't you, you can't smell that gas? But the smell of the gas, not long after eventually went away; and so did my concern about.

But other smells that were produced, had a range, from sweet fragrant perfume type smells, to the foulest types of unpleasant stench. And at times when I smelt some of the more unpleasant smells, for instance: a smell like someone had shit their pants or something, I had moved away from where I was, or mentioned or complained about it – when nobody else smell it. And, as had been proven to me, this producing or mimicking of smells' was another one of the Entity's many abilities – and which it seemed to have used for its own amusement.

Something else it had been doing to me for many years also; was making me overly hot, 'raising my body temperature: often to the point of perspiring; and generally in places with other people around. But no one else seemed to be affected in the same way. This, another of its abilities I had now also been made aware of; and that had been proved to me through further demonstrations, that it had also been responsible for. And which over the years from time to time had also baffled me.

So, not only was I glad to hear about this support group for people who heard voices, but was curious, about what these other people were experiencing. And if they too, were like me; possessed! And if they knew it! So the following week I called the number and arranged to start going to the group; and which met on a Monday afternoon, and was located at Ashley Road near Archway in North London.

The Entity had already told me a while back, not long after it had revealed itself, and proved to me that it had also been responsible for the symptoms I had thought of as schizophrenia, 'that other people experiencing the same symptoms accompanied by voices in the head; were not ill, but were possessed by a spirit. So to some degree I already knew and was aware of this. And this, as well the unusual story of how I had come about this information, was the main reason that I had wanted to write the book; to enlighten the world to this bizarre reality. And now a few days later! Now the Monday: I was about to meet some of these people. Though during this difficult and often frightening week,' my main motivation to go to this group had been because of what I still having to deal with; the threats' and the not knowing if this Entity, was still going to carry out its threat to snuff me out in my sleep. This was still causing me a great deal of anxiety, and for me the group at this point, was to serve as a crutch; another source of help for my wellbeing; somewhere to go; and where I could air this and listen to what others had to say; and also, find out if what the Entity had said about them being possessed was a 100% true. And again, on the day I went there for my first time, I decided to not say I was possessed by an evil spirit and that I knew what I knew. But again, say what I had said at the Haven' how the voices had been terrorising me etc.

The group itself – and sat in a circle. Consisted of no more than six people, including the group counsellor-facilitator, who was in charge of the group and set the format. First by talking in general, asking if anyone had any problems and was o.k. and if anyone would like to ask anything, which sometimes they did; and to which others in the group, would then add their comments or experiences. And what would follow this, later he would ask each person how things were with them, or how the voices had been or were – allowing for a ten or fifteen minute slot, sometimes longer depending on what the person had said. This too, would sometimes' or often raise comments or input from others, commenting on having the same or similar experiences, or having to deal with the same type of thing or situations.

And it was this focus and this format; of having each person have a ten or fifteen minute slot,' to say how they were, that was the most revealing! And (as the weeks passed by) making me eventually realize, that they 'were', all "possessed;" just like the Entity had said. And after going to this group only five times; I noticed that although it was a small group, usually no more than five or six people. "That it was not always the same ones every week! There were some regulars: but different people came and went. But, what struck me the most,' was the different degrees of suffering they were having to endure. And even more harrowing was that they were ignorant and totally in the dark as to the real cause. And it wasn't just the voices they talked about and how they were affected by them, but all the other symptoms that I had experienced were also present – the paranoia, the delusional behaviour, hallucinations; all contributing to their overall misery

The threat by the Entity several weeks earlier, which had led me to this hearing voices group; and its postponement to kill me on the following Friday obviously had amounted to nothing! It saying to me the day after that eventual day, that it had now postponed it for a year,' and that I now have a year to live! And although I knew the evil' this Entity was capable of. I also knew it liked to play games – nasty and evil ones, it had a mischievous side. And, while I couldn't be sure, a 100% that it wouldn't carry out its threat; for the time being I did my best to ignore this latest postponement: putting it down to its game playing evil nature! And instead, decided to concentrate on ways of getting through the days, or finding ways, or places to go, which would take my mind off things. Basically try to limit its influence on my life.

"Only a couple of weeks earlier, the second week of November, I had had my last and final appointment with the Psychologist at the ear hospital. This had been a good source of help. And, although I hadn't intended the appointments to have gone on for over a year,' it was because of what I had been going through with this Entity that I had kept them up, as during the difficult times I found just going there was helpful, and in some way, therapeutic, or beneficial; and something to do. And the appointments during the last six months had only been once every month anyway. And for these reasons hadn't mentioned that my ears had been fully cured since February; or the unusual story behind it all; or how they got like that in the first place. And in the end, had let the psychologist take full credit for getting my hearing back to normal: her thinking it was the coaching along the lines of psychological and practical ideas' that had in the end cracked it.

I was still going to the white eagle lodge regularly, and continued going to the Haven during the evenings, but now not as often, maybe three or four nights a week. And in my continuing efforts to try and free myself from this Entity; although I had already seen one Vicar before and which had been disappointing, another thing I had done was phone a Vicar at a local church: Saint John's on Holloway road; to see if he could help me. I had told him about the situation and what I was having to live with; and he told me to come to the church at a certain time on the Sunday, which I did, and spoke to him for a while; and he said some special prayers. He also told me to read the New Testament; saying it was part of a prescription he had for me. He was familiar with 'spirit possession, and spoke about it in a matter of fact like way. Saying he had come across cases of it in Nigeria where he was from. He also said I should attend the church service which I was now doing: attending the Sunday morning service. And I had started to read the New Testament. 'And after getting through a good chunk of it,' was surprised at how many references it had made in regards to possession by evil spirits; and what these spirits had been doing to their unfortunate victims; and it elaborating on how 'Jesus' had cast the spirits out of them. And although I was more than familiar with many of the quotes and stories of the New Testament, and the story of Jesus, I had never actually read the whole thing right through myself. And although in the past; 'I had heard references to Jesus dealing with evil spirits, and, was even sceptical about the existence of them: these stories now took on a whole new meaning! But another thing that had struck me about it was how the authority and power of Jesus shone through in word and deed: saying to a person in the church that I had talked to on a couple of occasions, that it was powerful stuff, to which he replied, 'yeah, it changes lives'.

On Monday afternoons, I continued going to the hearing voices group at Mind. I had now been going there for two months. And although this Entity was generally mischievous, sometimes to the point of evil and I had been put through those terrifying couple of weeks because of it. It was around this same time, about six weeks after I had started attending the hearing voices group' that it had begun to reveal more detailed, and specific information to do with the spirits that possessed these people.

This became a series of revelations' which first began one night while I lay in bed; and after, I had been thinking about the people in the hearing voices group, and then trying to go to sleep. And continued in this same fashion over the next four months: First, beginning by telling me that there are different types of Entities and Spirits in the universe. But the ones that possess people and are responsible for the voices and the other symptoms that accompany them, and what is largely regarded as mental illness i.e. schizophrenia, are all the same type of spirit; and are, a "Non Human-Spirit Entity." And saying that they all have the same abilities to delude, control, produce the voices, the smells, the hallucinations, heal, or cause physical ailments and illnesses. Anything it could do they could do also, and that it also' was one of these – this same type of Spirit Entity.

It elaborated further, saying: but although they are all the same type of Entity with the same abilities and character, which is basically mischievous, evil, and highly intelligent. They also have their own personalities, which are as diverse as the personality's of the people that make up the human race. And that it was this element that accounted for the varying extremes of symptoms from one person (or one victim) to another. Saying, that although the character of this type of Entity was the same in nature, it was the individual personality of it that accounted for the degree of suffering, or how the symptoms would present themselves in a person who had been invaded by one. For example, some will live in the body of the person and only cause minimal problems, allowing the person to live an almost normal life; but manipulating and using them in some way. And others at the worst end the scale will cause havoc and misery and will delude the person into doing the worst kinds of things – even murder. And the proof of this was not only in those possessed by these Entities and what they had to say, 'but it further elaborated by reminding me of how it had deluded me into doing things during the years that I had not known of its presence' and influence on my behaviour; and during the years that I had spent in the Psychiatric Hospital and had been diagnosed as suffering from schizophrenia.

Just one of the many examples it reminded me of, and was one of the more disturbing insights into their ability to delude and control a person, occurred one evening, about eighteen months after I had moved into my flat, and during the phase that I had saw improvement in my eyes and legs.

During this evening I had arrived home to my flat. I had just come back from the West End. It had been a pleasant day. I had had something to eat there, and had later wandered around Soho; and had wandered into the more seedier parts of it; and out of boredom and curiosity had wandered into one of the adult shops'. And looking around at the various stuff on display; I don't know what had got into my head, but I decided to buy one of these blow up love dolls; and took it home. Anyway, after arriving home, and now later in the evening I had put the box with the doll in to one side; then about an hour later had decided to blow it up to see what it looked like. I did this and then stood it up in the corner of the bedroom, leaving it there; and then turned the light off and went back into the living room. It was now getting late' and about an hour later feeling tired I had gone and lay down on my bed. This was something I had often done to rest my eyes, and had become a bit of a habit. I had not intended to go to sleep; but had dozed off; and then woke up about an hour later. 'The bedroom light I had left off,' and the bedroom had been slightly illuminated by the light coming through the hall from the living room; creating a silhouette type of low light in the bedroom.   I lay there on my back, enjoying the calm and peacefulness of night; listening to the silence, and pondering over whether I should just close my eyes and go back to sleep. And in no hurry to make a decision, had then noticed the purchase I had made earlier, the blow up doll standing up in the corner on the other side of the room. I moved my head so as to have a better look at it; and noticing how much different it looked than before; the dim level of light making it seem almost like it was real. My attention then drifted away from it, and for a short while other thoughts had occupied my mind. But then I looked at it again, and had imagined, or from what I now know, had been deluded into thinking that it was really alive; even becoming convinced that I had seen it move; and that in some way it was planning to attack me if I dozed off again. Gradually these thoughts became more intense, the doll becoming more real, the dim silhouette of light, clouding and obscuring its features, and it appearing to me like it was playing a game, and just waiting for the right time to strike. Finally: becoming totally convinced that it was going to strangle or stab me the moment I went to sleep.

Now totally deluded! I got up from the bed, not looking at it or giving any indication that I knew its secret, and went into the kitchen, got a steak knife from the draw and concealed it behind my back; my thoughts now angry and worked up into a frenzy; outraged at what it had planned to do to me. I went back into the bedroom and approached it as fast as I could. Throwing my walking stick aside, and bringing the knife from behind my back, plunged it into the doll some twenty or more times, stabbing and slashing at it. And not convinced that that was enough,' went into the bathroom, filled the bath, deciding to drown it as well; holding it under the water for a good ten minutes. And just in case that there was any chance it would come back to life again, I got a bin-bag and put it in there, tying it, and wrapping other bags over it, then took it downstairs, and outside, putting it in a rubbish bin.

This was just one! But there were many more such incidents that had occurred throughout the years; this one being at the more extreme end of the scale. I was reminded of others, which it had been responsible for, the Entity producing them in my head pointing them out. Another one it reminded me of, was my threatening to throw a good friend through a top floor window, after having been deluded into thinking he was plotting to steal something from me and becoming intensely angry about it: This one happening years before the other one. 'These two incidents,' were just some of the delusions at the worst end of the scale 'that it had caused. And are a good example of the power and ability of these Entities' to delude and control a person.

The next revelation came two weeks later, during the evening and in the same fashion as before, in the dark, while in bed. During the day I had gone to the hearing voices group at Mind; and had heard one of the people there saying that sometimes she heard more than one voice in her head, saying that she had heard two voices talking about her, and that they had different accents. And I had wondered if maybe she was possessed by two of these Entity's. And it was on this night while in bed that the Entity demonstrated the ability to produce more than one voice. And produced a conversation with two voices in different accents, and male and female, and what were very different characters; thus, giving that impression. And saying to me it was just another one of their abilities, used to aggravate, or play games, frighten, or just generally annoy the person.

But then it went on and explained how the voices were produced! Saying: that what happens in the case of the voices is that this specific type of Entity is able to break into the body of a person; and inhabit the same space as the human spirit. (What we may term or regard as the life force 'or the electricity that animates the body). Thus creating a situation where the functions of the body can be animated by it in the same way: Almost like a Siamese twin situation where there are two heads and more or less only one body, but both able to operate it and have the same control over the body in every way, in every aspect. The example being, two spirits in one body: the human spirit who the body belongs to, and the invading alien Entity who has gained access to it. And the explanation for how the voices are produced' was explained as this: The same way that we can all think and talk in the privacy of our own mind and hear our own voice, talking, thinking, and resonating in our head etc. This alien Entity puts in motion this same function. It will think and talk in this same way; using the persons brain to do so. And thus: the result is that the person hears voices in their mind, in their head' and knowing full well, that it is not theirs, not their own: And the voices that they produce in this way, are clearer and sharper and generally louder than the sound of the persons own voice resonating in their head, and are therefore, automatically distinguishable, as being something foreign. And when first heard by the person can be quite startling or frightening.

And they can do this in different accents and at different volume levels and intensity's of sound, whether quiet or louder or faint. And as well as what was later demonstrated on me, also create an echo effect and also project them so they seem as though they are coming from outside of, or around the head: And are therefore able to use all of this to create all kinds of scenarios and fears. And the mischievous and evil nature of this type of Spirit, makes it so this situation is not predictable; and with the voices ranging in nature, from appearing to be nice, to mischievous, to nasty, to evil, or even helpful.

This process of how the voices are produced, also explains the reasons that the speech centre of the brain has often 'been seen to be being activated' when brain scans, or the monitoring of brain activity have been done in studies, or tests, to try to establish what is happening in the persons brain, that is experiencing this problem with the voices. And that's why this speech area can sometimes be seen working, being animated' during these tests, because 'this' process is being put in motion: the possessing Entity is using the persons brain to produce the voices.

The Entity elaborated further, by saying that whether the voices were produced or not produced during these studies and tests would have been down to the individual personality of that Entity. And when this speech centre is seen to be activated during these times, it would be the result of the game playing mischievous nature of this type of Spirit. It also said a similar process that is used to produce the voices is also used to distort and delude the persons thinking and perception, this time using the thought processes, and brain chemistry, to affect the 'behaviour' in all manner of ways, manipulating or influencing or deluding the individual.

The stuff about the speech area of the brain often seen to be being animated during these studies, was first brought to my attention several months earlier by the Psychologist at the Ear Hospital'. As a

154

result of having the label or diagnosis of schizophrenia this got brought up and led to her asking me if I heard voices (which I said I did) and which we talked about a bit; and which led to her drawing me a diagram of the brain and speech area: in her effort to try and convince me that the voices were the result of a brain malfunction of some sort. This during a time when things had been particularly bad and she had asked me what I thought caused the voices, and I had told her, that I thought it was a spirit: But didn't tell her that I knew for sure it was a spirit.

The Entity also clarified something else! And that was that this type of Spirit Entity is not in the Aura of the person, like some people had told me. And which I had also read things about them being in the Aura – 'on the outside of the body.' But said that was laughable, and using my own voice and shaking my hands about, said: do I feel like I'm in the Aura?

I pretty much well knew that anyway. And when I had been told that the Entity was in my Aura, had not been fully convinced – thinking to myself that this thing is not in my Aura it's in my body. But being fully aware of its range of abilities I had also kept an open mind, and didn't fully dismiss this possibility either. And now also because of the development of specialist equipment' the Aura can now also be photographed. And this seemed to be further confirmed, when sometime later and out of curiosity; mainly that I might see a hint of its presence in some way, I had my own aura photographed to see what would show up. But there was no sign of the Entity: Nothing that would be recognized as an abnormal presence! But just the colours of my Aura; which was mainly violet, light violet, 'with some yellow.

But although this was not the case! It did seem: by the range of things that it could do and had been done to me, particularly those appearing to come from outside of me and around me, that it must in some way be able to alter, or project, or expand itself in size in some way for these purposes: somehow being able to expand beyond the confines of the body?

Another reason it had clarified this was because of the belief thought by many working in the field of spirit removal that all of the spirits that possess people, are human spirits! And that they attach themselves to the persons Aura; and are in the Aura' of the person, and operate from there. But it said this was not the case; and certainly not with this type of Entity.

'And when it confirmed this to me, I asked it if there were human' or other types of spirits that did attach themselves to peoples Auras. And also, what other types of spirits were out there, "were their Angels and other such things. But its response was: I will only tell you what is relevant to you.

I also asked it during these times about the spirit world at large and what was out there, what did it look like. And again it refusing to tell me anything; saying only, that it could tell me all kinds of things about the universe, but you wouldn't know if I was telling you the truth, and once again, saying that it was not relevant to my situation.

I even asked it if it had a name, or parents, and being a bit sarcastic, brothers, sisters. And also what life span these Entity's had if they had one; and had it possessed other people before me, "which I was pretty sure it had. And also what did it look like, its appearance, and what form did it take, in shape etc. But again did not get any straight answers, being told it wasn't relevant.

"Something had certainly changed during the weeks and months following the terrifying ordeal, that had lead me to the hearing voices group at Mind.' The Entity had subtly worked its way back, not so much, into my confidence; but to a level where the situation was much more bearable. And during the nights while I lay in bed and it was revealing these specific details, it was often a pleasant and fascinating thing to experience. Not just the information I was being told. But the whole tone of this Entity had become very serious, and took on a quality that showed a side of it that I had not yet seen. And even the tone of the voice it would produce would have a very serious and noble quality. And combined and blended with its other fascinating abilities to flash thoughts into my mind and project images or moving images into the minds-eye, I was often captivated by the uniqueness and magical brilliance of it all. And besides these more specific points of information that were being revealed to me, it would also elaborate or comment on some of the people at the hearing voices group. On things they had said, 'telling me more details about this type of spirit. And it also said something else! And that was that the information that was being revealed to me in such detail: that no one else knew about it to this degree, and I was the only person in the world to have been told this; and my situation, 'was unique.' And that what I had experienced; and because of what I had experienced' and what had been revealed to me, I was very privileged.

To some degree I was already aware of this, but I certainly didn't feel privileged. I'd thought about what I'd gone through. But I certainly knew what it was revealing to me regarding this question of schizophrenia and the voices, that there was no doubt about this. I had now seen enough to know this was the absolute truth. This conclusion I had reached after about four months of going regularly to the hearing voices group; and listening to the various people that had been unknowingly invaded by these Entities, and what they said, the stories they told; and the general information that was expressed. And as far as being privileged, I suppose in some ways I could say I was; but this more a result of the whole experience from when it had started; the amazing healing abilities, the methods of communication etc, the uniqueness of it all; and being enlightened to the reality' of invisible beings at large and amongst us.

It was now the case that new information was being revealed to me on a regular basis. And had now been four months since these 'revelations' had begun, and now well into the New Year – now March 2002.

I had continued going to the church regularly, almost every Sunday; and continued to read the New Testament; and spoke with and kept the vicar informed of how things were, usually having a good talk with him after the service and which culminated in him saying special prayers for me. Overall things did seem better. But as well as the information that had been revealed to me. In between some of these better times the Entity still continued in its little ways of aggravation. But as I sometimes learned, and thought what had been aggravation; later turned out to be another way of it revealing something else? One of these being the ability of these Entities to induce their victim into a sleepwalking state, and have them do things without them knowing about it.

What I had noticed during and around this time in my flat, were marks on the walls, and gouge marks on the doors. And had wondered where they had come from; because I knew for sure that I hadn't done them. And on the days after noticing these, other things would appear: gouge marks on my new bed and holes ripped into my clothes. It was then, soon after, that the Entity made me aware that it had been doing these things; by inducing me into a sleepwalking state and using my body to do them. And as more proof, it further elaborated, by reminding me of something that had occurred a couple of years back. When after getting up one morning and going into the living room I saw a chair had been broken and was lying on the floor. Baffled as to the cause, and especially so because my back at that time had been in such a bad state, that I knew that it could not of been me that had done it. And although I had thought this at the time, the only other possible explanation I could reach was that I must have done it in my sleep.' And it was this incident that the Entity now reminded me of, saying it had done it, it had broken the chair. And its way of proving it could do this to me – and display another one of its abilities: was to do these other things, and show me them the next day. It reminded me also, of the screenplay, and bits of it that had gone missing; 'the one scene I had rewritten,' and other pieces of it that had mysteriously disappeared: Telling me that they had been flushed down the toilet! But, this didn't seem to be enough, and for the rest of this week it continued to damage other things. Whether this was to further prove this beyond doubt or just be a nuisance' I don't know. Other things it did were: cutting the stitching in the seams of my trousers, and making a split in my bankcard, and which when I repaired it with super glue would split the repair the following night. And what I had started doing because of this was to check my clothes before I put them on. Check them for holes or rips or if seams had been cut. And not only was it disturbing to know it could do this but even more so because I had absolutely no recollection of it, none at all! And during this time;' after waking up, wondered what it had done today.

This also answered the mystery as to why I had sometimes woken up to find a bruise or scratch on one of my arms or legs. Also baffled as to where it had come from. But what I found the most distressing, was that the Entity was now flashing images into my mind of things it could make me do, without my ever knowing about it, i.e. like going outside and attacking someone, or a whole range of other things that could get me in trouble. And this was something else that played on my mind, and that I now had to think or worry about also.

"And something else" that had recently fully dawned on me! Was the ability of this Entity, to influence and manipulate my thinking; to the degree that I now realized' that I didn't even know my own mind? This was something else that had gradually been revealed to me in a number of ways: its ability to influence me without me realizing it.

This was something I was already aware of anyway; but mainly in the capacity that was used to delude. But I was now also being made aware of a more subtle use of this ability, which was used just for everyday purposes.

An example of this being: When I had first mentioned, shortly after the magical whirlwind had occurred, and during when the healing operations and exercises had first started, that I felt like I was being guided, and seemed to automatically know what to do next -- like the thought had been fed into my mind! And later, when this was confirmed, 'by the thoughts becoming bolder, and having a feel that accompanied them.' 'It was this same form', or ability, that these Entities have, that can be used, in different phases. The Entity telling me, that this could be done in three stages – like a switch, low, medium, or high. The low form, so subtle that the recipient possessed person is not aware of it. The medium form, being more noticeable and having a slight energy and feel to it, suddenly coming into the mind and quite recognizable. And this was the one, that the first stage had evolved into and I began to distinctly recognize, during the weeks that had followed. And this third, high stage, this was much bolder and is instantly recognizable; and the one that is now generally used, being used as another way of it talking to me or commenting on something. And this was also the same with the images being produced in the minds-eye; these too could be produced in a low, medium, or bolder high setting.

I remember during the time that I had been writing the screenplay, often great ideas or dialogue would suddenly just come to me, appear in my mind. Some of which was often brilliant and fitted perfectly, and which I was often inspired by what I had thought was the creative process at work, and even wondering where it had come from as some of it just did seem to pop into my mind: almost as if by magic. But this was something else the Entity had recently told me – that it had often helped me with the screenplay: using these more subtle forms to do this. And it told me it is this first stage, this low subtle method that is the one that is used, on those that are possessed by these Entities, to influence and create all kinds of problems for the victim, by mildly manipulating their thoughts and decisions 'without them ever knowing about it. And I don't necessarily mean to always delude the person either, but more so to have an input, or stop the person from doing things, or to generally cause disruption to the person's life. Or even influence them into not doing things, or doing things they really don't want to do, encompassing a wide range of areas, from how they react, to even what they say: even affecting mannerisms. Or just simply make them act out of character: "generally interfering with their free will!

And it was this aspect of the ability to subtly manipulate and even control using this method that I had recently become more and more aware of, gradually realizing, the overall effect that this Entity had on every aspect of my life. To the point when fully realizing this, an even greater realization dawned on me; and that was: that I didn't even know my own mind. And the Entity confirming this reality; saying to me, that if it left me, I would be introduced to myself.

And it further stated that it is another aspect or variation of this same subtle process that is used by these Entities to delude the people possessed by them, into doing the more terrible things that are sometimes heard on the news or read about in the newspapers: 'often the person has said voices in my head told me to do it. But the truth about this aspect and what the Entity specifically stated was this: That it is not the voices that make them do these things. Simply put, a voice in your head telling you to do something doesn't mean that you will go and do it. And it is not the voice' that makes them do the more terrible things. But it is the ability of the possessing Entity to apply these subtle forms, combined together with an ability to delude the persons thoughts and perceptions that is behind it; winding up the person, causing paranoia, delusion, or manipulating them into a frame of mind where they are convinced that they are right in what they're doing or seeing, and making them act on the delusions, and therefore carry out a terrible deed, attack or murder someone or commit some other type of heinous act. The voices are not necessary, not at all: as I knew only too well!

' During the times that it had deluded my mind and perceptions into doing some of the more extreme things, I had never experienced any voices, but never the less had been deluded into doing them. It said in these cases that it was just down to the personality of that Entity whether it was accompanied by voices or not: whether deluding the person first then additionally goading them with the voices. Or just using the voice to say things and then manipulate and delude their thoughts and perceptions. Saying to me that the voice in the head on its own would not make them do anything, and it was these subtle controlling delusional abilities that were 'first' put in motion. Bringing the person into an obsessed and deluded frame of mind; and that it was this, the use of these delusional abilities that were the sole driving force. And that they could flick this on and off like a switch, for whatever lengths of time; using it in this way in different extremes on a daily basis, to manipulate, control, or toy with the person.

The other point the Entity made clear and that has to be highlighted also: As well as the explanation for how the voices and other things are done or produced, is the other element which it couldn't explain: The supernatural element! The other abilities it has, and how they were done; the hallucinations, the smells, the different methods of communication, the ability to heal anything in the human body, in the ways that were done to me; or cause illness and health problems. Or some of the more bizarre things that I had experienced, like the whirlwind energy or the electrical net; and also the ability to produce and project the voices, so that they seemed like they came from outside the head, around it, or very close to it. It said, overall, these were things that it couldn't explain how they were done, and although to us they were supernatural, and things we found hard to fathom. It said these abilities' were just part of its makeup, and were natural to its state, and which all of this type of Entity has at its disposal and can effortlessly put in motion in the blink of an eyelid.

It seemed strange to be being told and enlightened to the reality of what is not mental illness, but something far more harrowing; to most people almost something incomprehensible. And although this Entity had been revealing the finer details of this reality and the real horror story that was behind it, which I had seen and been put through myself; and that their seemed to be a different personality emerging from the Entity. It was still often putting in motion its aggravating, annoying, or mischievous side – not as much and not as often, but it was there. And with this in mind, I knew, overall; long term it wasn't to be trusted. I had even wondered if it had just revealed all of this information to me to just goad me; and make me fully realize just what amount of control it had over me, and that it was influencing me with me having absolutely no idea about it? And that maybe it was just one of the more particularly evil ones. And thinking about everything else it had put me through; as well as the years of hell with my physical health. And it was these things! Things like this I would ponder over, knowing full well just how evil it could be.

And, as far as any progress on the eventual removal of this Entity and my continued efforts on bringing this about, I didn't really feel that all these things that I had been doing were going to work, or be successful. It just did not seem like it was going to happen. I wasn't being negative, but brutally honest with myself; thinking that I could still be here in a year's time doing all of these same things and still, nothing! I was trying to be patient, attending the church every Sunday, and giving it time. But after four months of it I was increasingly becoming more impatient and disillusioned; and starting to feel the same about the healing at the white eagle lodge. And my disillusionment with going to the church was further increased after a recent conversation with the vicar.

A month earlier I had given him a copy of the long letter that I had written for Graham Wyllie. It was over twenty pages long, and explained the whole story in more detail. My intension in doing so was because when I had first started going to the church and had told him about the extraordinary healings and what this Entity was capable of, he had said very clearly that this spirit couldn't heal me of anything, and only God could do that. I did get a bit annoyed by him saying that, but at the time couldn't be bothered trying to convince him otherwise; and instead, had let him revel' in his self-righteousness. And as I have already mentioned, was sick of repeating the story anyway. And all that was really important and that he needed to know: was some basic information. And that I was possessed by an evil spirit. And more importantly for me – could he get rid of it.

I did however, about two months after this, mention this long and 'more detailed' letter I'd had to write to this person, in order that it may help him in his efforts to remove the Entity: 'It explaining this whole situation as it had occurred. This during a conversation about the many people I had already seen in my attempts at removing the Entity from me. And to which the vicar said he would like to read it, and the following Sunday I brought a copy with me and gave it to him. And by the next Sunday, after the service I had spoken with him and he said he had read half of it. But after reading only half of it, I think he must have probably thought that I was mentally ill: saying to me, 'I think it would be good if you see the Psychiatrist'. See if you can try some different medication or something.

My immediate reaction was disbelief, 'and are you serious, or are you joking? But he was serious; he certainly wasn't joking.

He had already asked me before I gave him the letter if I had a history of mental illness. And to which I tried to explain that I thought I had been mentally ill, but now knew it was this Entity, that had been the cause of what I had thought had been mental illness. And, as he had asked for it, had let him read the long letter so he could see how this whole situation had developed and just what this Entity was capable of: And my also thinking it was additional information, that may in some way help him in his efforts to remove it. But it seemed instead to have had the reverse effect. And after that, on leaving the church that day, I decided not to go back again. Thinking to myself, what's the point? As I had walked along the road after leaving the church, I thought to myself, "that I wish this Entity could possess him for a couple of weeks; and do some of the things it does to me on a daily basis; he wouldn't have any doubts then! I know damn well he would have then said, 'yeah' I see what you mean. But anyway, it wasn't the first disappointment, I was now getting used to it. But something else I had been doing recently; weeks earlier before I had left the church I had started going to a Sai Baba Centre in a part of London called Mill Hill. This was the person in India Margaret Palmer had told me about and I had written a letter to the previous year – the so called miracle man.

Anyway, this Centre had been difficult to get to. It was also on a Sunday morning, the service starting at 8.30, and I had gone there before the church service. And although I had been there three times already, it had been a real struggle to get there; mainly because the Entity had been stopping me from sleeping. This was another one of its abilities that it was now also revealing and had emerged around this time, and which I was now becoming familiar with: the ability to produce the effects of insomnia, making it impossible for me to fall asleep. And, if it wasn't producing the effects of insomnia', another thing it would do to stop me from sleeping, was jerk my head or body, stopping me in this way, usually when I was about to doze off. It played around with the both of these.

The insomnia by itself; was another one of its horrible skills, that had recently emerged: And another reason why I was right to continue the pursuit of getting rid of it. And more horrible due to the fact that I had often used sleep as a form of escape: from what I was feeling, around this time; was an unpredictable misery. And during many of these nights that I had been kept awake I had only managed to get 3 or 4 hours of sleep. One of its patterns was, it would let me fall asleep then after an hour or two wake me up and keep me awake for hours on end; then allow me fall asleep again, then after an hour wake me up again; knowing full well that I was planning to go somewhere the next day. It was deliberately ruining the situation for me so I would feel awful, or not go at all.

The service at the Sai Baba Centre only lasted an hour or so, and after which I had gone to the church service. But it was not long after I had started going to this Centre, some weeks later that I was told about a Sai Baba Centre that was in central London – Russell Square; and that the service was in the afternoon,' starting at 4 p.m. which was much better for me; and as I soon became aware after going there twice, was also more suitable and likable all round.

The service in both of these Centres had a format that centred around the singing of Bajans (devotional hymns) but firstly began with the chanting of a mantra; then the Bajans would follow lasting for about an hour. But what was different about this centre in Russell Square was that towards the end after the service had finished, there was a speaker who would get up: usually someone different every week. And talk for 15 or 20 minutes, sometimes longer. And either talk, about Sai Baba's teachings, or read passages from his teachings, or sometimes combine the two; and which were often profound and inspirational. But more importantly: they would tell stories of personal experiences of Sai Baba's miracles; and which not only inspired me; but gave me hope as well.

Chapter Eight

The Avatar

Avatar, or Purna Avatar. Meaning: fully divine incarnation. Or simply put: God in human form. This very notion, or idea, to most people would be hard to believe or even comprehend; and likewise, in this respect, I too was no different. But Purna Avatar means just that: God in human form! And it was this that Sai Baba claimed to be --- the Avatar of this age: A modern day Jesus.

It was around this same time, and not long after I had started going to the Sai Baba Centre in Russell Square' that I had finished reading another book about Sai Baba. This had been the second one, and had been written by an English couple who had had much personal experience of being around and observing Sai Baba at close hand; and had written their own accounts of many personal experiences; and of miracles and supernatural phenomena that they had witnessed: 'over many years.' And for these' and many more reasons, they had been left in no doubt of what Sai Baba refers to as, and terms his omnipresent, omnipotent, and omniscient attributes. These being: the qualities of being present everywhere, of having unlimited power, and of knowing everything: All knowing, all seeing, and all hearing. And of which Sai Baba says are all natural qualities possessed by the Avatar. Things you would attribute to God! And it was these powers, that I had continued to hear and read about, in the many stories of how he had answered prayers for his devotees, and confirmed his remarkable abilities to help people; often in very unusual circumstances; and who were in all corners of the world, while he himself was in India. And all they had done is prayed to him by saying his name, or praying while looking at a picture of him.

And these remarkable powers and abilities that were attributed to him, and which ranged from: the miraculous curing and healing of all manner of things, the materialization of objects; as well as all kinds of help in many forms; to knowing' the past, present, and future of everyone. And even well documented raising of the dead. And something else: His teachings and incredible wisdom and extraordinary knowledge. 'And it was all this; as well as being fuelled by my own situation; that had gradually provoked my own curiosity to find out more.

With many people on first hearing about Sai Baba, and some of these Stories; they would initially invite disbelief; and in this respect, I too was no different: In fact, even more so. This because of the variety of people I had already come across in my attempts to rid myself of this Entity, and who promised things that they didn't deliver: and, all of who' had been confident that they could get rid of it. So in this respect I was still wary of the claims of certain people to have the power to do this and that.

But the situation I was in was also different, in that I was someone who himself, had experienced something very unusual and unique and knew because of it that these miraculous abilities, supernatural abilities, to heal and do other things 'were' indeed possible, and a reality. So in that sense 'for me' it wasn't a question of 'were the things I had heard and read about Sai Baba being able to do, 'possible.' But, could 'he' do these amazing things. Was it true? And was he genuine! And it was questions like these, I had sought to clarify; and which had prompted me to find out more about him. And not only that, but the harsh reality, that he now seemed to me the only one person who could genuinely help me; I didn't know anyone else! And also now didn't particularly have any faith in anyone else. And, had also come to the stage, where after having read two books about him, as well as finding out there has been 'at least a hundred books written about him; and many more volumes of books, filled specifically with his discourses. 'Combined with the other sources of information that came my way, while at the Sai Baba Centre. I had now reached a point where I genuinely believed; based on the collective evidence, that this person, 'Sai Baba', whatever he was or claimed to be, he was indeed, something special, to say the very least.

He himself has said that he hasn't come to start a new religion or solve all the world's problems. But to restore the ancient highway back to God; to teach the fundamental truths about human existence, and to avert humanity from catastrophe. And by his tens of millions of devotees worldwide, a figure that is ever increasing, his work or reason for being here is certainly having an effect. And I suppose a testament to this man being something very special would also show in the type of people who are his devotes; and which has a range of people from all walks of life, from the poorest to those who hold the most responsible or highest positions in society. And as I gradually found out, many of which would have become devotees not just because of his teachings which themselves alone are often fascinating and inspiring. But because they too, would have put him to the test in some way, either through prayer or another form, or had been the recipient of a miraculous healing, an answered prayer, or several of them; or had directly been witness to and experienced his omnipresent attributes, or some other extraordinary happenings.

So the many reasons people had been drawn to him were not just because of what they had heard or read about him: but concrete results from direct experience. And as Sai Baba himself has said: "do not try to fathom or understand me, but experience me. Meaning in other words put him to the test and see what happens. And as he himself has said, that there is no problem he cannot solve. And that he can simply will something to happen! And it is also interesting, as to what he describes God as too.

At this time of writing this he is 76 years old. And although most people in the west have not heard of him, there are also many that have heard of him, and are also devotees – and growing in number. A 'devotee' being those who generally live by a nine point code of conduct' that is specified by Sai Baba. But he himself does not advertise himself: Saying' that the true Avatar needs no advertisement. But also says he draws people to him for a particular reason. And something else he has said is that he will live here on earth until the age of 96, and then leave his body. Also saying,' that only the 'Avatar' can know his time of death.

96 is a ripe old age to begin with; with only a small percentage of the population ever living until this age. So reaching this age in itself is an astounding feat. And I could see no point in making such a claim if he couldn't live up to it. Why bother; it's not necessary. To not fulfil this would only have made him look not all that he claimed

to be. There would be no point. And there were also other things I had read about him that were extraordinary: one of them being the Aura that emanated from his body; and it not being the Aura of a 'human being. "This first noted by a researcher! Who had a scientific involvement with something called Kirlian photography? (Biometric field radiation photography) first developed in the late 1960s, in the former Soviet Union. And which when photographing a person with this, would also show up the normally invisible Aura that emanates from the body: "Scientifically known, as the bio-magnetic energy field. And just for reference an Aura of a human being usually expands or radiates on average anywhere from one to five feet from and around the body. And which, can differ in colour from person to person; and often have a mixture of colours, or, one colour in particular being dominant. And' just like the more than six billion people in the world: no two look exactly alike. And even certain parts of the Aura can change colour during times when affected by emotions, such as anger, fear, stress, affection, depression, hate etc: with the differing emotions' each reflecting a colour of their own. And now, as I mentioned in the previous chapter, as this technology has been further developed, people can now have their Auras photographed, as well as have an analysis giving meaning to the various colours. And it is widely accepted that certain colours in a person's Aura, are associated with character and personality traits or certain gifts, each colour having a special significance of its own. And the Aura itself, also comprised of energy bands, also has another purpose, in that it is a protective barrier. Something I will talk more about in the next chapter.

Anyway, this researcher and scientist, his name: Dr. Frank Baronowski, and who was based at the university of Arizona, had originally got involved in the research with this Kirlian photography; because he himself had possessed a gift, ' in that he could see peoples Auras; and had watched and observed them since childhood. A gift very few people possess. And it was this that had lead him into this field. And with him later wanting to further his research, during the 1970s had travelled around India, and for scientific purposes had visited and observed and photographed a hundred or more holy men, with the intension of discovering if any of their Aura's had anything special or different about them. He had found nothing that was significantly different, only that some of them had slightly larger Aura's; 'but nothing to write home about. Until he arrived in Puttaparthi, home of Sai Baba, and had observed his Aura, and to his astonishment could see that the Aura of Sai Baba was not that of a human being (his words). The Aura itself stretching and projecting thirty or more feet in all directions; as well as it having other attributes' that were also unique.

And, so now, for me, it wasn't a case of being a gullible mug, and simply believing and hoping, but it was this catalogue of evidence from the many different sources, conveying, that this 'Sai Baba, had been put under the closest of scrutiny. As well as what I had also heard and read myself! To now be convinced enough, that overwhelmingly; he appeared to be genuine; the real thing: 'something special and worth trying to seek help from.' And as I've already said, he now also seemed like the only person left who was able to help me. And now I was no longer going to the church, going to the Sai Baba Centre had by now taken over from it. Now I was going to the Centre almost every week. And not only was I participating in the service, but while there, I had also started to pray to Sai Baba.'

It seemed strange, to be praying to this person. But I had come to the conclusion that it was worth trying, and what's-more; I had absolutely nothing to lose by praying. And it didn't cost anything. And it seemed to be one of the easier things to do. And shortly after I had reached these conclusions; in the weeks following, I also decided that as well as praying and going to the Centre: 'that I would also travel to India in person to seek his help. And then set about finding out the next best time to go. Because as well as there being literally thousands of people at his Ashram (spiritual community) at any one time; there are also times of the year when it is even busier – these being when certain festivals or celebrations occur'. And then, after much enquiry, I had decided to go in September, which at this time was about five months away. Being told September would be very quiet and not too busy; and also wouldn't be to hot either. Plus it gave me enough time to organize and get the funds together.

Meanwhile: while I waited and planned for the day to arrive that I would travel to India, I was still having to endure the unpredictability, unpleasantness, the mind games, and sometimes cruelty that the Entity inflicted on me. But overall, I had been learning to get by and live with the situation; and was being patient; and working towards my next attempt' at getting this Entity removed from me. I still thought about the idea of writing the book and telling the story, and even more so now due to the specific revelations that had been revealed to me. But knew also, that for the time being that for now, while this Entity was still with me, I would have to shelve the idea! 'Until the time arose, when which it was no longer with me. And so, I hoped that the trip to India to see Sai Baba would be successful, and that at some point after my return; I could start the process of writing my unusual story.

And, as well as my planning for the trip to India. I now also realized because of the situation I was in, that, I would also need to have somebody to accompany me! And the main reason I had decided that this would be necessary, 'was because I had recently got a passport, a couple of months earlier, and now the Entity had been threatening to do something to it. Saying it would do this while I was asleep; induce me into the sleepwalking state and then either rip it up or throw it in the bin, or damage it in some way. And knowing how evil it could be, I couldn't afford to let this happen, or take the chance of dismissing it as trickery, mischievousness, or not taking it seriously. After all, my passport was my passport to India and Sai Baba, and so I had to act on this; and I took it very seriously.

A couple of months later: It was now July, and only two months until September. And, I had now acquired the funds for my trip to India. And had also solved the problem of what to do about keeping my passport safe! I had found a suitable person to accompany me on the trip, and who would also keep hold of, and look after my passport. In fact not long after the Entity had made its threats to do something to my passport I had given it to this person for safe keeping. This was someone I had met the previous year, while at the sports hall in Crystal Palace, when I had gone to see 'Mother Amma. And as I had been leaving the building I had asked this person for directions on how to get out of there. It turned out that she also was going back to central London and had a car, and said if I waited for half an hour she would give me a lift. Her name was Juana, she was Spanish, and from what I later found out was a Sai Baba devotee; and had been for about five years. And, during the journey back we had talked about health matters and things of a spiritual nature - including Sai Baba! And had later exchanged phone numbers and kept in touch. And it was her, who had later told me about the Sai Baba Centre in Russell Square and had first taken me there and introduced me to the place. I hadn't at first told her about the Entity; only that I was having to live with an unusual problem and that that, was the reason I wanted to go to India to see Sai Baba. But as I sometimes bumped into her at the Sai Baba Centre, and sometimes spoke to her over the phone, eventually I told her the bizarre story.

She as it turned out had already been to see Sai Baba once before and was also planning to go and see him again: sometime before the end of the year.' But didn't have a confirmed date planned and was herself saving towards the trip. And so she knew all the things that I needed to know, about the journey, and other useful bits of information. She also later borrowed me a book about Sai Baba: this called 'Sai Baba Man of Miracles'. So I was now on my third book about him, and finding out even more. And sometime after I had told her about the Entity, and my reasons for wanting to go and see Sai Baba, and knew of her desire to go as well, I had later said that if she wanted to go sooner, and accompany me on my trip to India, and look after my passport the whole time, I would help her with the cost of the trip – which she agreed to. And during late July we started to make plans and settle on a date for the journey.

I had already been told that after the 8th of September to the end of September was the best time to go. Then after looking around at certain travel agents had finally got a confirmed booking and date to fly on. The date was the 11th of September' and we planned to stay for two and a half weeks, until the 29th.

One thing I hadn't at first been aware of, and that had recently been brought to my attention; was that simply going to India to see Sai Baba, did not necessarily mean that you would get to speak to him. I had read this in this book. And Juana had also told me first hand, of the procedure' and how Sai Baba operated: Telling me, that he walks along the aisles among the people, in the Temple at certain times, and picks out people for interviews. And that was how you got to speak with him.

I had already read that he does this on the basis of knowing the past present and future of every person, and knows why they have come to see him. "And generally picks out people most in need of his help. Whether this is physical, mental, or spiritual help. And when he is referring to the past, he doesn't just mean this life, but what he would term past lives. And something else! Although he has cured many people of many things, he did not always cure everyone, even the ones he had given interviews to. And his reasons for this he says; was the person's karma from a past life.

Karma was something I had heard a lot about during the past few years; and was getting to understand more about what karma was. But as of yet, had not fully understood its significance, or, the finer details', of what it was all about. 'So my going to India was no guarantee of anything!' But nevertheless, I remained positive, feeling that I had nothing to lose and would go anyway. In fact, I had no choice. There was nothing else, no one else to turn to, I had done everything that I knew of, there was nothing left to try. And in the meantime had begun to make preparations for the trip; buying necessary items, getting a visa, and the necessary injections and malaria tablets.

The weeks went by; and the day finally arrived; and off we went to the airport, boarded the flight' and were on our way to India.

'The first thing I noticed after takeoff was that the plane was half empty. I had actually been told by the travel agent that they were the last two tickets available. It had been hard enough trying to get a flight for early September and I had already tried other travel agents. So I had been pleased I had got them close to the day that I had been advised to go. But as I had looked around me I had noticed that the travel agent had obviously lied. It was the 11th of September; the first anniversary of the attack on the twin towers in New York; and this had been the reason for the plane being half empty. Most people when booking their flights had obviously avoided travelling on this day. And I can't say I blamed them either, as I wasn't overly enthusiastic myself. And the truth was, the travel agents were probably having difficulty trying to fill seats for this day. And it was for that reason that I had been lied to. But as it turned out, the lack of passengers had actually made the flight more pleasant and comfortable all-round. My only complaint being the lack of sleep I had had throughout it. And there seemed to be a heavy security presence at the airport,' and from what I gathered, extra vigilance. So in reality it turned out to be a good day to have travelled on.

After arriving at Mumbai and changing planes, then flying on to Bangalore, we then travelled by taxi, for some three and a half hours to Puttaparthi: the home of Sai Baba; taking in the scenery, and many sites along the way. Juana had already made arrangements for a place for us to stay in, booking us a rented apartment not far from the ashram, which she had stayed in last time. The journey had gone smoothly, and finally, I was here.

During' the previous days, that had lead up to the actual day of the journey; I had thought a lot; and pondered over the often strangeness of life; and had also gone to the Sai Baba Centre on the Sunday: three days before the trip. And as I had sat in there, I had thought back to the very first time I had come across Sai Baba. It was when I went to see the Psychic Surgeon 'Steven Turoff' in Chelmsford. "It had been while sitting in his waiting room." I had noticed a big poster high up on a wall; 'with a picture of a man,' who I would later come to know' as Sai Baba. And recalled' that it had said in writing along the bottom "Baba". It caught my attention! This man dressed in this long orange robe; and I remember having a good look at it and thinking to myself: what's he supposed to be. But I also noticed a mirror on another wall with words written on it, saying, 'Baba loves you.' And there was a written sign next to it saying, don't touch this, 'Baba materialized it.' And I remember thinking at the time, that it looked as though it had been written with red lipstick; and to a large degree had been somewhat dismissive of it. But as I had sat there in the waiting room another part of me had later thought – and mainly due to the nature of the place I was in, and the things I had heard about Steven Turoff, "that maybe these materializations are possible, who knows?

But that was my first encounter with Sai Baba. And at that time I didn't even know his name was Sai Baba. The only name that had been on display was 'Baba. And I didn't enquire about him or anything after that, and had completely forgotten about him; until two years later, when Margaret Palmer had mentioned him, and his miraculous abilities; in view of the fact that I might want to seek his help, telling me about him and then showing me a picture of him, and which I recognized instantly, telling her what I had seen in Steven Turoffs waiting room. But at that point and time, when Margaret had told me about him, it was about six months before the magical whirlwind had occurred, and, was in no state to travel to India. And furthermore I wasn't convinced that he was entirely genuine either: This mainly because of the many people who I had already seen, the various healers and the like; and which I had been told and convinced, that they too may be able to cure my health problems. And so, when Margaret had first mentioned him, and told me some of the things that he was capable of, I had, to a large extent thought that it was an exaggeration', and that there was probably something bogus or phoney about him; and had not been that interested in what she had said. But now, here I was a few years later, in Puttaparthi, India; and now overwhelmingly convinced of his powers and seeking his help, and about to set eyes on him. And it was this unpredictability; and strangeness of life; which I had pondered over during those previous days.

It was early in the morning when we arrived in Puttaparthi, about 8 a.m. and after being shown the apartment and getting settled in, I just wanted to have a sleep, and which I did soon after this, and sleeping for most of the day. The apartment itself was spacious and very nice. It was a two bedroom: each having our own bedroom and bathroom; and with a ceiling fan in every room.

The first thing I had noticed and had really struck me about being in India was how hot it was. And although September was supposed to be one of the cooler times of year, it was still very hot. And after having had a good sleep late into the afternoon; later during the early evening I went out for a stroll. Had a look around, and bought some provisions. But it was the next day' that I finally got to see Sai Baba for the first time in the flesh. It was in the afternoon during his second appearance of the day and what is termed and known as Darshan' (him appearing and giving his blessing to his devotees).

This was in the Temple, which is huge, and when full can accommodate 20,000 people or more. With men sitting on one side, and women sitting on the other side, and both having separate entrances. And as I noticed during this time, the Temple was about half full.

The Temple itself was pleasant and tranquil, the main part covered by a roof. There were no walls, obviously because of the weather, only a brick and wrought iron type wall of about six foot high. And which had a stone type bench that ran right along the inside of it, throughout the length of the Temple; and where I had initially sat. But the normal procedure unless you were sick or disabled in some way, was to sit on the floor in organised rows.

And to see Sai Baba for the first time seemed quite strange. I hadn't seen him arrive, as I had got there a bit later. He was sat in a thrown like chair which was on a raised platform and about forty feet from the front of the crowd. And there were from what I later gathered, students from the local collage sat at his foot and who were engaging in conversation with him.

But besides this, the first thing I had noticed about him was the robe like gown he had on: And more specifically, the colour of it. Or even more specifically the shade of the colour. It had struck me straight away; this because I instantly realized where I had seen it before. 'It was in a dream I had had about three weeks before coming here.

The dream at the time, although bold, and stood out, and leaving an impression on me, and with me remembering every aspect of it, was significant in the sense that I had read, and had also heard stories from other sources that Sai Baba often comes to people in their dreams. Often to give a message, or sign, or some other information; or to just make the person aware of his presence in their life', and that he is aware of them: giving them some reassurance or something.

And to me, because of what I had learnt, as well as everything else I had read about him, these dreams were just another aspect of what he was capable of. But even when I had had the dream; I had, because of the abilities of this Entity and knew the types of things it could produce, first thought that it was either just a bizarre dream,' or

176

that the Entity may of been responsible for it. And so to a large degree, had given it no importance and took it as a pinch of salt.

The dream itself was brief, lasting probably no more than half a minute. But powerful! It was Sai Baba standing in a tree in the distance and he was saying something to me, and I couldn't hear him, or make out what he was saying. But apart from this, what really stood out was the boldness and richness of all the colours, and in particular the colour of the gown he was wearing. It was orange, but not the shade of orange that I had been used to seeing him wearing in the pictures I had so far seen of him. In these, he always seemed to be wearing the same colour every time – a lighter shade of orange. And what was different about the gown in this dream, was that it was a darker and deeper shade of orange, much bolder, vibrant, and noticeably different; almost contrasting the difference between a pale blue and a royal blue, instantly noticeable; and a colour I had not yet seen him in. And as well as all the colours in the dream being rich and vibrant, this one had stood out especially; and had been something I had distinctly remembered. And it was now while here in Puttaparthi and on first seeing Sai Baba and noticing the colour of the gown he had on, that I saw this colour again, and instantly recognized where I had seen it before. It was the exact same shade of orange as the gown I had seen him wearing in the dream; everything about it. It was a perfect match. It struck me to a degree that I had at first had a few double takes, and then, looked again at it throughout the time that I had sat there, just in case I was wrong; or checking again that it was the exact same colour; which it was in every way, the same as the one I had seen in the dream: it was a perfect match a 100%. And after I had sat there for an hour or so, and had later left and walked away, as I was leaving the Temple, I had looked back at it once more. But by this time, had started to think, that maybe it was just a coincidence' that it was the exact same colour; and had more or less settled on that theory. But as I was to later learn; this was no coincidence.

Over the following two days, I had familiarized myself with the routine of the Temple; observing Sai Baba, and more specifically, where I needed to be if I was to stand any chance of gaining an interview, or even talking to him. And was soon informed' that it was overwhelmingly in the morning, at morning Darshan that he gave interviews. And, that if I was to stand any chance of getting one I needed to get up at 3 a.m. to get there by 4 a.m. and then wait patiently sitting in organized lines outside the Temple: Until the time came when around 5.30 a.m. the person at the front of each line, picked a numbered ball out of a cloth bag, and then each line would get to go into the Temple in the order of that number. This was significant, in the sense that when you went into the Temple, you were then guided by staff where to sit. And this starting' right at the front of the 'aisles' that Sai Baba would walk along: when he came out and took letters off people, talked to people, and picked people out for interviews'. And the aisles themselves were wide and spacious, about 20 feet in width. And that,' if I was to stand any chance, of at least getting to talk to him; then I needed to be at the front, in one of these first three rows.

So now I had established and found out what I needed to do, I set about putting it into practice. Starting by first buying a small cushion for me to sit on, and which would make the whole thing a bit more comfortable during the long waits. Then the next morning got up at 3 a.m., got there for 4. And on this first day of doing this, I got placed in the second row along one of the aisles. And which, during this first time, Sai Baba didn't come anywhere near me. The second morning was the same. But it was on the third day that my luck changed and my line number came up higher in the order, and I was placed in the front row of one of the main aisles.

I was excited and pleased that I had managed to get the front row. And sat there; waiting patiently for Sai Baba to appear. And which at 6.45 a.m. his usual time to appear, like clockwork, on the dot, he did so. Then his usual routine would follow. First he would walk around the female side, talking, taking letters, and picking out individuals or groups for interviews. And after which he would then proceed over to the male side of the Temple and the same procedure would follow. "And the ones, who had been picked out for interviews,' would then go over to another part of the Temple: to sit and wait outside the interview room; until he had completed his walkabout.

I had now learnt enough and knew enough about Sai Baba, and the procedure of obtaining an interview, to know that by simply asking for one, would not mean that you would get one! As I had read about and had even seen people asking him for one' and him not granting them one. And you certainly can't buy an interview or offer him a bribe or money, it doesn't work like that, this would have no effect at all. As I had already mentioned and been informed, apparently he knew why you were there and knew who to grant interviews to on that basis. So this was something I was well aware of. And knew that if he came near me and I asked him for an interview, that, 'it did not mean that I would get one. The general rule was that you waited to be asked by him. And with this in mind and my intension of sticking to the protocol; and somewhat keeping in mind that he would know why I was there, this is what I intended to do: simply wait, be patient, and see what happens.

So I sat there in the front row, patiently waiting, waiting for him to finish his walkabout in the female side; and observing his movements. When finally after some time he finished and came over to the male side and began his walkabout here. And I knew for sure by the aisle I was sat in, that he would at least walk past or come near me, as this was one of the main aisles that he always walked along. So not only was I in the front row, but in a good spot too. And although I knew the procedure regarding an interview; I had done something else that I thought might help in my obtaining one. And this was in the form of a letter I had written and put in an envelope, and which I had in my hand, 'ready to hand to him if he came near me. And the content of which basically explained and told him, that I was possessed by an evil spirit and that I wanted him to get rid of it. And although overwhelmingly,' I had been convinced of Sai Baba's abilities', I had written the letter anyway, as an addition, just in case, 'covering all angles; doing as much as I could to make him aware of me and my unusual problem. And I suppose also within me there was still an element of doubt; that maybe he didn't know about me and my situation.

So anyway, he eventually made his way down the aisle where I was sat; moving along on the other side of it, and was getting closer to me. He then took some of the letters from people on that same side of the aisle, but by now, was almost directly opposite me. Then he did something that caught me by surprise. He all of a sudden turned right around and walked straight over to me. "And before I knew it he was standing right in front of me, and looking down at me.

Initially I was stunned', and although I wanted to say something; just held out my letter, which he took from me. He then asked (quote) from where have you came? And to which I replied 'England', and couldn't think of anything else to add to that. "I was genuinely lost for words and still trying to gather my senses: trying to think of something else to say' or keep the conversation going. Almost like I was dumbstruck and hoping he would say something else. It was then that someone behind me, put their hands out, raising up their palms towards him; and that Sai Baba then moved his hand in a circular motion, and materializing what is known as and called vibuti (sacred ash) and sprinkled it into this man's palm. I then raised up my palms, putting them together in the same fashion, and received some of this vibuti. Which as well as it having a symbolic and spiritual significance,' is often given by him as an aid to cure health problems. People either rub some on their forehead or eat it, or both.

I had watched his hand as he had moved it in a circular motion, and had seen this vibuti appear out of thin air, almost like magic, and had been impressed. But at the same time it was something I had heard and read about too: also reading many accounts by devotees of it manifesting on a picture they had of Sai Baba. And therefore accepted that it was something he could do; and so, wasn't overly' impressed by it. But, before I knew it he had been and gone, moving on further down the aisle. And although I had been pleased that he had come over to me, and had felt a sense of euphoria for the rest of the morning because of it, and had been glad he had taken my letter; the regret of not saying anything more to him had started almost immediately after he had left. Thinking to myself, that here I was presented with a golden opportunity, and that, "I could have at least just asked for an interview anyway. And why didn't I think of that at the time – as my situation was an unusual and desperate one. But I later consoled myself by being positive, and thinking that there was still twelve days left and that I would persevere with the same routine; and maybe, I would have another chance. And also knowing that he had taken my letter', and that this in itself was progress! And also people I had spoken to after and mentioned that he had come over to me, 'had said that that itself was a real blessing and I was lucky.

The days following this first encounter' with Sai Baba went by quickly. And with me continuing the morning routine and hoping that once more I would get the front row again. And as well as this I would also go to the Temple for afternoon Darshan almost every day too. And, in-between these times, or late afternoons when I had free time, I would either rest, or look around the many various shops in the surrounding town, or go for a meal somewhere; or sometimes go for a sightseeing tour on one of the motorized Rickshaw vehicles that were used as taxis. And being driven around in these three wheeler vehicles' was also a great way to cool off. The main tour to see; was all things to do with Sai Baba,' and covered a large distance lasting about an hour or more. And which the rickshaw driver would point out, or would stop outside for a few minutes or more so you could have a look or take a photo. And the main things to see on the tour; were: where Sai Baba was born and grew up, to the many things he had built, from collages to the huge state of the art hospital, these providing health care and education to the poor and less well off free of charge. And although I had already been on the main tour once, I found myself almost every other day flagging down a Rickshaw vehicle and asking to be taken on a tour again. And the main reason for this: was to cool off – relief from the heat. The breeze and wind flow that the almost open topped Rickshaws generated was a superb way to cool down. Also kill some time; and enjoy the scenic landscapes of the surrounding countryside.

On one of these tours we had stopped on the top of a mountain, and the driver had pointed to a river in the distance, and at the same time telling me the name of it. And as I had got out of the vehicle to have a look I had said, man it's hot.' 'And his reply' was that this was not summer. So I could only imagine what that was like. This for me was hot enough. Sometimes I felt like I was in an oven. And I suppose the view of the river would equally sum it up, as the river he pointed out was bone dry, and at first thought I was looking in the wrong direction, until he informed me.

Another driver who had taken me on a tour had later taken me to his house and introduced me to his family. The home itself' was no more that a small shack amongst an enclave of shacks. I had been a bit cautious about accepting his invitation. But out of curiosity to see how the locals lived decided I would go along; but keep my wits about me. And after arriving there and being introduced to his wife and kids and being given a cup of tea, I was then what I would call, subjected to a monologue of stories by him and his wife; and which were intended to arouse my sympathy into handing over a fixed sum of money. And one of the main reasons he told me they needed this money was that he wanted to buy his own Rickshaw vehicle, as the one he was driving he paid rent for. I told them I would have to have a think about it to see if I could help them out, and said I would look out for him at the Rickshaw Rank, the following morning at a certain time after the Darshan; and let him know if I could help. I had thought about maybe giving them a donation; but certainly not what they were asking for. But later that evening, after giving it some thought and not yet fully deciding what to do, the Entity butted in with some good advice and very accurate observations. And telling me not to bother and not to give them anything: "Saying that it was a well honed double act that they had done many times before. And to further convince me and being even more specific, had highlighted certain things about them and things they had done; and the way the act had unfolded.

I had during the time while there, had thought and got the impression that there was something staged about the whole thing anyway. And the Entity being what it is and not missing a trick, had only further confirmed this. And so in the end I took the Entity's advice, and didn't bother' to look for him the next morning at the Rickshaw Rank – and left it at that.

The Entity up until this point had been well behaved, and had not caused me much trouble at all. It had been very quiet. And although I didn't ask it, I had wondered how it felt about being here; knowing full well my intended purpose for being here was to get rid of it. And as I was well aware that it knew everything I was thinking, this meant that it knew also that I had been thinking about this. But yet, it had not yet commented or butted in, like it sometimes would during times when I had been thinking about things. And although I could have asked it; a part of me didn't really want to know; or even care. But, it was times like this, and this kind of help, which was one of its good sides,' and which I had gradually become accustomed to; and was, becoming more of a feature. It would warn me about things, give me good advice, and tell me things about people; making excellent observations. And it was during these times that I thought,' that if it could be like that all the time, it would be an asset and a great thing to have, and the fact that it could heal anything in the human body – who wouldn't want that.

It had been a 100% accurate about what it referred to as a well honed double act, and the specific things it had highlighted' further pulling apart their act. And this was all something very valuable, that in truth, I knew I would miss when the time came and it was no longer with me.

Many more days had now passed by. And although I had kept up the routine that I hoped would gain me an interview, or just the opportunity to talk to Sai Baba again; as of yet it had not brought this about. And the days were now drawing closer to our allocated day of departure, now in the beginning of our final week here. And as well as the efforts' I was already making I would also pray almost every day too; usually in the Temple after the afternoon Darshan. Hoping it too would have some effect. But, as the day of our departure drew nearer, I was gradually becoming more pessimistic about my chances of gaining this opportunity. Now with only four days to go.

But, it was on the fourth morning before our last day that Sai Baba had done something else that had surprised me. I had got the second row on this particular morning, and had sat there as usual waiting for him to appear, and in my hand had had three letters. One was from Juana and one from her friend in London.

Juana had been trying to get Sai Baba to take these letters for the past two weeks, but had not been successful. And when she had phoned her friend in London and had mentioned this, and had also mentioned that Sai Baba had come over to me etc. And as it was the last week her friend had said to give the letters to me, feeling that maybe I would have better luck. The third letter was my own, and which had been written on the advice of a woman there who Juana had got talking to and was knowledgeable about spirits and possession etc. This woman had told me to write a letter – a letter detailing some specific things, and then hand it to Sai Baba. And although somewhat reluctant to take her advice and write the letter; I had done so anyway, feeling that I had nothing to lose by writing an additional letter.

This whole business about Sai Baba taking a letter from you is in itself symbolic, as he himself has stated that he already knows what is in the letter. And he doesn't always take letters from everyone who hands them to him either. So when he does, it is supposed to be a good sign. 'I had personally witnessed, on numerous occasions, him walking by people who were trying to get him to take their letters. I'd even seen people putting letters just a few inches from his hands and him not taking them. There was one particular incident that stood out from the rest. This, when Sai Baba had stopped to take some letters and a man had stretched forward across three rows of people and had pushed his letter within an inch of Sai Baba's hand; and he didn't take it, he snubbed him. I remember thinking', that was a bit mean. But him taking a letter is also symbolic to the individual; in the sense that he took it and in some way approves of them and that they may be granted what they have asked for. To the devotee it's like a blessing: something quite special. And, had by now, read and knew enough about Sai Baba to understand its significance.

And now, as I sat there once again, waiting, letters in hand, and wondering what today would bring; the time passing, and getting closer to the time of his appearance. When as regular as clockwork, I heard the familiar sound of the quiet gentle music that would start to play; and signal his arrival, and play throughout his walkabout. And once more I watched the routine, with him first going to the women's side then gradually making his way over to the male side, then, him gradually making his way down the aisle where I was sat.

He had stopped at the beginning of this aisle and attended to some people; then continued along the aisle, but was now walking directly in the centre of it - on the red carpet; and I could see he was getting closer and closer. Then suddenly he stopped: and now about twenty foot away from me. He then turned, facing me, and looked directly at me; and looking me directly in the eyes; obviously clearly making a point or a beeline for me for some reason. But what he did next was strange; and at first wasn't sure what was going on' and was a bit puzzled. As he had stopped and faced me and directly looked at me, he then started saying something to me which I couldn't hear; almost like he was mouthing it to me. I at first thought that maybe he was asking or was going to ask if I wanted an interview, something of this nature.

This spectacle went on for about fifteen or twenty seconds; and which seemed a long time for someone to be standing there doing that; and very odd. And I couldn't work out what was going on, 'or what he was saying. He then stopped this and then walked on, and then came within about five foot of me: taking letters from the people in the front row, just to my left. It was then that I remembered the letters I had and stretched over placing them near his hand, and which he took from me; but didn't look at me, as other people were doing the same thing. And then like before, before I knew it he had moved on; slowly making his way down the aisle.

I was bemused by the spectacle that had just occurred, and also somewhat disappointed, as I had almost felt that he was about to ask me to come for an interview. And fearing that with now only three days left' it did not seem like this was going to happen. But I had heard of him picking out people on the very last day of their visit, and kept this in mind; still feeling that there was still hope.

It was the next day, during the early evening that I had gone to a cafe for something to eat, and as the cafe had been busy a woman who couldn't find an empty table had asked to sit at mine. 'She too was British, and we soon got talking. She was a Sai Baba devotee and had been here once before; telling me also, that she had had an interview with Sai Baba. I wasn't feeling to good at this particular time, and was actually feeling quite depressed. The conversation eventually got round to what had made me come to see Sai Baba. And, not really caring what she thought of the unusual reason for my being here had said that the reason for me being here was due to being possessed by a bad spirit and that I was trying to get rid of it. I elaborated further by telling her more about it: just in case she may have thought I was a nut of some kind. But as the conversation went on, and by her reaction, and behaviour, and what she said, I could tell that she had taken what I had told her very seriously. I had also told her later of the spectacle that Sai Baba had performed the previous morning. And telling her I couldn't make sense of it. To which she had said after giving it some thought, that maybe he was talking to the spirit – communicating with it in some way. I was met with a similar reply when I had told the woman that Juana had introduced me to, the one who was knowledgeable about spirits and who had told me to write an additional letter, her saying this also; that he was talking to the spirit. And when she had said this same thing,' I thought maybe he was. I thought that was probably it, and thought what else could of he been doing. It seemed the most plausible reason and also the only one that made any sense; thinking that maybe he was giving it a warning or something – as bizarre as that maybe: 'and what a carry on this all was. But after that, and for the remainder of the stay, I had put this to the back of my mind and gave it no more thought.

"This same woman, had also later given me some written information; an article on how Sai Baba had removed-cast evil spirits out of people' who had been possessed, and had come to see him.

But anyway, the day of departure soon arrived! Our scheduled time to be picked up by taxi and taken to the airport was five o'clock in the afternoon. And as it was the last day, I had once again during the morning gone to the Temple. And on this last day; Sai Baba did not come anywhere near me. I also went there again for the afternoon Darshan, giving it one last try, just in case; and also, 'to give me the peace of mind that I had pursued my goal right up to the end. But as I sat there during that afternoon; and knowing it would be the last chance I would have; and after Sai Baba had came out and gone through his usual routine, and then went and sat in his chair – a signal that his walkabout was over. I knew then that that was it and that gaining an interview and getting rid of this Entity, for now, for some reason, and which I didn't know; was for the time being not meant to be. And, that soon I would have to leave and say goodbye. And as this reality fully sunk in, and took hold, a deep sense of 'depression and melancholy came with it.

In my view I had done all I could and there was no more I could do. And looking at my watch and realizing it was now time to go, I got up and slowly walked out of the Temple for the last time. And as I walked past the outside wall I had a last look back; 'one last glimpse of Sai Baba. Then a slow walk back to the apartment. And further sinking in also, was the full reality of the unpredictability that lay ahead. Now what was I gonna do I thought!

I then tried to put the disappointment out of my mind; and arriving back at the apartment organized and assembled my belongings ready for the journey ahead. And an hour or so later, the taxi arrived, and we were off. And for me, the three and a half hour drive to Bangalore Airport was for many reasons an unpleasant one. The main one being that I was travelling back with the Entity. And who once again had now turned very nasty! And for most of the drive to the airport' had been taunting me about the fact that I had not succeeded in getting rid of it.

It was doing this by using the voice in the head. And as well as the taunting' was also saying evil and nasty things to me, or about me.

It started this not long after we had left Puttaparthi, and didn't really stop until we got to Bangalore. And after it had first started, it had really brought home, and reminded me once again, of the full horror and madness of what I was having to live with. The gradual effect' was a numbing one that sent me into a daze. "I looked at the others in the car. Observing them, 'peacefully viewing, and taking in the pleasant sights and scenery of the Indian landscape; and it was probably peaceful thoughts that had occupied their minds. And it seemed bizarre, that here I was sat in a car, and being driven through the Indian countryside, with two other people, and was being verbally abused by an evil spirit. How bizarre was that I thought! But this was my reality; and I did my best to ignore it. Letting it carry on with its vicious verbal tirade.

After eventually reaching Bangalore and boarding the flight to Mumbai and then changing for the flight to London, there was a sense of relief that we were now on board for the last stage of the journey home. And the Entity' had gradually tapered off its nasty behaviour, not long after we had reached Bangalore. And although I had a lot on my mind and was feeling low, I also felt calm and relaxed; and the flight itself was generally pleasant. But it was about two hours into the flight, that I had once again, thought about the spectacle that Sai Baba had performed to me a few days earlier. And once again had pondered over its meaning: or, if there was one at-all. When all of a sudden it struck me like a lightning bolt. 'That what he was actually doing was; he was mimicking the same thing he had done in the dream that I had had about him three weeks before I came to see him: "The one where he was standing in a tree in the distance and was saying something to me that I couldn't hear or understand; and wearing the rich vibrant orange coloured gown. The exact same colour gown' that I had seen him wearing the first time I had seen him in the Temple. "I then worked it out: 'What it was all about?

I had mentioned earlier how when at the time of having the dream, that although it was powerful and left an impression on me and that I remembered everything about it. But knowing' what the Entity was also capable of hadn't taken it seriously; thinking also, that it could have been the Entity who had produced it. 'Even though the timing of it was significant; in the sense that it occurred three weeks before I came to India.

'But, I had still not fully endorsed it, as being a message, or a sign that had come from Sai Baba.' But now it all became clear; crystal clear. He had put together this specific dream for me; knowing I was on my way to India, and that he would be able to not only now make me realize a 100% that it was him who had been responsible for the dream. But also, to make me fully aware 'that these amazing things,' that people said he could do; were really true, they were real. And he had done this, firstly: by letting me see' on the first day I had seen him in the Temple; the exact same coloured gown he had been wearing in the dream! "But, by him knowing even after I had first seen him in the Temple 'wearing this same coloured gown' and it being a perfect match to the gown in the dream,' that I had not been fully convinced, that there was a reason for this coincidence 'and had later more or less dismissed it. "And the reason being" that he, 'Sai Baba' had produced the dream!

But, as I had said earlier, as I would later learn; this was no coincidence. And by him knowing what I had thought about this coincidence and that I still had some reservations about his abilities or omnipresent powers, and even the ability' to know what you're thinking, that on this particular morning he had singled me out to further convince me in another way. This, by mimicking the same thing he had done in the dream. Him suddenly stopping in the aisle, on the red carpet, turning, facing, and specifically looking directly at me and mimicking exactly what he had done while standing in the tree – in the dream; mouthing something to me, something that I couldn't hear' or work out what he was saying. The exact same thing!

Obviously his way of saying that it was him' who had produced the dream? But also to say: if the perfect matching gowns were not enough to convince you, that it had been me who had produced the dream, will this help; and now do you believe it was me. But the more I had read and got to know about Sai Baba and understand the way he does things, and the way he can do things. 'That's Sai Baba through and through.' He doesn't always do things in a way you would expect. And I suppose, knowing that he had acknowledged and went out of his way in such a way to prove this to me; and also proving he was genuine, and also aware of me. This was a good sign. Making me aware that at least he now knew about me; and that maybe he had some kind of plan for me.' And after a lot of soul searching; this was the only conclusion I could reach?

And so, as I sat there on the plane, the mystery of the spectacle now solved. It not only left me fascinated and in awe of just what Sai Baba was capable of, but also feeling that all was not lost. And as I had said earlier, that maybe for now, for some reason, I had to live with the Entity. I certainly needed time to think and reflect, and plan my next move. And instead, of dwelling on this during the journey, had instead, put it to the back of my mind. The rest of the flight was a pleasant one, and seemed to go by quickly, with us eventually touching down and landing back in London. 'My immediate thoughts' were that it was nice to be back on the ground; and back to a cooler climate. And on later arriving home' thinking also, that it was nice to be back in my flat. But also a sense of feeling: now what?

Chapter Nine

The Noble Task

Yes: now what? These were the sentiments that echoed through my thoughts; the uncertainty of my whole life, and the bizarre situation that it was. And the weeks following the arrival back home to London were a mixture of ups and downs; depressing and frustrating; and with me often thinking and regretting the time Sai Baba had come over to me. Regretting' that I had waited for him to ask me things, and regretting that I had not just boldly asked him for an interview. Or as Margaret Palmer had later said when I had told her of the encounter, 'that it wouldn't have done any harm to have said when asked by him, "from where have you came", to have added after saying 'England', 'and that I've come here because I'm possessed by an evil spirit! And the way she said it made it sound so simple and easy. And made me think even more, and with regret; thinking, yeah, I wish I would have said something like that. And why didn't I? Thinking that there would have been no need to ask for an interview or wait for him to ask me. I could have just simply just replied with something like that.

But that didn't happen. And as I've already said, I was taken by surprise, and at the time lost for words. And it was easier said than done. And besides, another reason and theory I was coming round to regarding my situation, was that for now for some reason maybe the Entity was supposed to stay with me. And if this was the case' then Sai Baba would have known this. And for that reason had not given me what I wanted. And maybe for the time being he had just been acknowledging me. This was one theory I had settled on, and a theory, which was now top of the list of possibly only three theories. And although top of the list; it was also a theory that came and went, back and forth, throughout the weeks that had followed my return.

But there was a lot of information that I was now putting together and which further convinced me that this main theory was probably right. Or that it just seemed right; with also many things pointing in this direction. And another part of this main theory, was now the thinking along the lines, that maybe the reason that I had experienced and been put through what I had, and what had been revealed to me in the way the whole thing had unfolded, and the uniqueness of it all; was because, that maybe I had come into the world for this purpose: 'to fully experience and reveal in specific detail, the truth behind what has largely been thought to be a mental illness.

And reflecting on the whole process of how everything had unfolded and the way it had 'occurred, in stages, also suggested to me that it had been done in a measured and very calculated way: suggesting that a lot of thought and planning had gone into it. And other things seemed like further signs of this. Like the Entity telling me, the day after I had got back from India; and while I was about to take the malaria tablets, that have to be continued for four weeks after leaving there. 'Telling me in no uncertain terms', I did not need to take anymore: saying, "you will only get ill if I want you to."

And another area of thought that had further contributed to this growing main theory was that I had also, during the last year, been reading a lot about karma and reincarnation. And as well as already having heard past references to it, I was now engaged in exploring the teachings of it; by reading books that went into further detail about its finer workings. I had by now, after the bizarre and unique things I had experienced,' and was still experiencing, concluded that, although I would have at one time to a large degree, have been 'dismissive of my situation having something to do with karma and reincarnation. 'Now found that this wasn't something I could rule out, or dismiss anymore.' And now had to give some serious thought to, and at least explore the subject and see how it fitted in the greater scheme of things; and particularly with my own situation.

But apart from these more positive ways of looking at my situation, all wasn't well; and as the weeks went by, I had once again on two occasions reverted to seeking help. Visiting another person' working in the field of spirit removal. I had become more frustrated, and sick to death of having to endure the games or sometimes nasty side of this Entity, and was fed up, and just desperately wanted it out of my life. Because as well as it using its various methods to generally communicate things to me; these and its other abilities, were still sporadically put to use to aggravate, and toy with me as well - often in a nasty way. And the often frustration and anger that I felt because of this, and the depressing effect of it all, had at the beginning of December, led me to another person who apparently was able to remove spirits from people.

This person' had had a write-up in a magazine, which Margaret Palmer had seen and had sent me a copy of. It sounded quite good; and the way that she worked, was different from the other methods that had already been tried on me. And although sceptical, and still somewhat reeling from my disappointment in India, I had visited this woman at her home near Hampton court. And after arriving there and paying the fee and her going through her procedure, she had then told me that the Entity had now gone. And which soon after this, the Entity had confirmed to me, that it had not gone anywhere! And by it saying: of all the people you've so far seen, and methods used to get rid of me; this one, has been the most entertaining' and amusing yet.

But when she said it had now gone; and I was soon after made aware that it hadn't. I just kept silent, and let her continue talking and elaborating on the after effects, of a spirit being removed, and what I could expect to feel like. And feeling the sense of disappointment, knowing full well that it hadn't gone! I didn't tell her, that I knew for sure that it hadn't gone; just thinking to myself, what's the point. But another thing that had struck me' was that she had thought and had told me that all of these spirits were human – thus, making no distinction between any of them: 'And which I knew for a fact that this was not true. And only further confirmed to me that many of those working in this field do not know exactly what they are dealing with. And, as the Entity had later described: some of these 'methods used to remove it, as laughable. And saying also what I had already suspected, that they had no idea what they were dealing with.

"But although it had been a disappointment. Out of the continuing desperation I was still feeling, I had also gone to see her again a couple of weeks later, at a spiritual church she also practiced from. And which again she said the spirit had gone. Even insisting it had. Now saying there was nothing to take away, remove etc. Although this time I told her she was wrong. And questioned and challenged her methods: 'Which she didn't take too kindly to.

And so, this latest venture and attempt to free myself from this Entity, had only further confirmed and added more weight to the fact that no one was going to be able to remove it. And that there was only one person that it seemed could do this, and that was Sai Baba. And after this latest try I later gave up on the idea that any of these kinds of people could ever get rid of it. And had started to ponder on the question, of what I should next.

I had been back from India for over two months now; and although the initial first number of weeks back had been depressing, and low key, and with me coming to terms with the failure to of not achieved my aim in India; and the reality, of this now further sinking in. I had now, around this time started to take stock of the whole situation, and once again started to think and put together my next plan. And although I had already attempted during the previous year to start to write my story, and was stopped from doing so by the Entity; in terms of this unusual situation I was in with this Entity, this was also a situation that was continuing to evolve; and in the sense that I was getting to know the Entity better and becoming more aware of its games. And therefore took a lot of it in my stride. And had got to know that it also' had a serious side 'that was highly intelligent and clever. And by now, had gathered and become fully aware, that when the Entity wanted to display this side, and tell me a fact or something of importance, it also did it in a way, where that you knew, and was left in no doubt that it was telling you the truth.

And another part of this side that had continued to evolve, and had often become a valuable source of help: was its good advice. And this would display itself in terms of any everyday dealings, or problems that would crop up from time to time: 'it offering advice on things of this nature. And it wasn't that I just took its advice without thinking about it. But the advice it usually offered, was a better way of doing something or something I hadn't thought of. And realizing this and after thinking it through, acted on its advice.

There was often help in other ways too, with other abilities I was now being made aware of, and which it had put to use in helpful ways. This was its ability to see all around me. And which it would use to make me aware of things, often behind me – and which could be anything, and anywhere. And examples of this were: often pointing out things I was looking for, 'often in a supermarket, while looking round the aisles: 'And one time making me aware of a £20 note on the floor of a train station.

Another helpful way in which it used this was, when drawing money out of ATM machines, particularly at night; and when people generally tend to be more cautious and check who's around more – myself included. And when one time doing this; it had just simply flashed in my minds-eye the image' of an eye in the back of my head 'followed by a slight nod: Just to let me know it was looking out for me, keeping watch, watching my back.

It was also around this time that it had also told me, that over the years, it had looked out for me many times, and had protected me in numerous ways. Making me aware, it had been using this ability in its subtle form for as long as it had been with me. As well as it telling me that it had often been used mischievously and maliciously too: to make me turn and look at something, or someone - in a certain way, or look or gaze in a certain direction; 'reminding me too of some of these!

But it was this evolving process and this increasing role or input and help that it gave me, and also almost every time I wrote in my diary too, the help I received in this way, about the continuing unfolding scenario; and often many pages long. 'That it had now brought me to a point, where I thought, that maybe I could now start writing the book.' And maybe this time, instead of it stopping me; it may well let me get on with it? 'It was a gamble;' or at least something worth trying! And maybe' this is what I had to do before I could be free of this Entity. Maybe that was it, and maybe it was the intension of the Entity to help me do this so I would get it right. And maybe the reasons I had been stopped by it the first time was because it wasn't the right time to start it. And that there was still more I needed to know, and which would now also make sense. And I had even thought and pondered over the time a year and a half earlier when I had visited the psychic and ghost buster 'Terry 0 Sullivan' remembering what he had said about the Entity being a very powerful spirit. "And saying that he thought that it was not operating of its own accord, and that it had a contract with me: 'and until that contract had been fulfilled that it would not leave. And it was these words; that in my search for answers were now also resonating through my mind. And which now seemed that this could quite possibly be a reality. A lot of things were now pointing in this direction.' And maybe this was it – a contract? And maybe this was another piece in the jigsaw of my growing main theory. And that this was what I had to do before I could be free of it. "I had to tell the truth about the reality behind the voice hearers," and get it right; in specific detail. And that's where the Entity came in. And if I was right, Sai Baba would have been aware of this; and would have known.

But there was something else which I had recently remembered. And although at the time had baffled me', 'was now also starting to make sense." And also,' seemed to add more weight to this. And this was the time, two and a half years earlier' when I had gone to see a very gifted clairvoyant, called Lilla Bek'. Who had been highly recommended for her insight, and the accuracy of her readings. And for this reason, I had around two months before the magical whirlwind had occurred. And when my situation had been becoming increasingly more desperate, 'travelled to the south coast of England to see her;' in search of, and desperate for some answers. As by this point – as I had stated earlier, had felt that something seemed to be working against me. And was trying to find out if this was the case – and what was it? 'And, would there be an end to the physical hell I was having to endure; and hoping she could give me some answers!

The reading had lasted an hour, and was taped. And there were many points of interest, and also questions from me that had been highlighted and talked about. As well as a general reading, of what she saw was happening now and would happen in the future. "And said, that the reasons, I was going through the problems with my physical health, were for a reason, which would become clear, in the coming years.

'But it was something else she said towards the end. And that now', seemed to make sense! Saying that: (quote) the whole thing that has happened to you is not a huge gigantic awful mistake, it isn't, and that actually it is all for a purpose. And the main purpose is that, there will be a point in your life when you will do something really wonderful for humanity.

And it was this: 'that I would do something really wonderful for humanity! And although at the time had baffled me, and considering the state I was in thought it to be a bit ridiculous. 'But this, now, also seemed to resonate, and make sense.' And this: "doing something really wonderful for humanity! What else could it be? It wasn't something you would just say to anyone. And it now seemed to me, that this also, was another piece in the jigsaw; and which' was increasingly becoming clearer.

By now the Christmas season had arrived. I had spent Christmas in Wales; and while there, thought more about this. And in fact,' had planned to write the book anyway when I had got rid of the Entity. 'But this had not yet happened, and wasn't about to either. And, if I was right about my main theory;' then maybe this time, the Entity, would not stop me from writing the book, but may well even help me.

"The break in Wales", had also helped me break out of the often depressive rut or cycle, that had dogged me since returning from India. And I had returned to London with renewed vigour and determination: to embark on the next chapter of the continuing unfolding saga, that this had all become.'

It was now January 2003; and I had now been back in London for a couple of weeks! And although I had not yet started the book, I had decided that first I wanted to paint the kitchen in my flat; as it was badly in need of painting. And another reason for this,' was that it was where I also planned to write the book; as it was also the most suitable space to do this in. I also wanted it to have a change of colour, and a different colour scheme. And after looking at samples and shades close to the colours I wanted, I eventually made a choice; and set about the task of getting the job done. There was a lot of work to do. And as well as the preparation, there was replacement of tiles and some carpentry work and other bits. I got all this done over the space of about five weeks, and was completed by having new vinyl on the floor, new curtains, and a new table which was more suitable for the kitchen; 'and which I planned to write the book on.' I had done a first class job of it all, and the kitchen looked superb and sharp.

Other things that had been going on during this time was I had started going to the Sai Baba Centre again, usually once a fortnight; and had already started going again to the 'hearing voices group' not long after coming back from India; and either once a week or fortnight. And often found it to be a continuing source of help and somewhere to let off some steam. And it was from the hearing voices group that I had also been informed by the group facilitator: 'that there was a hearing voices conference', to be held in London on the 10th of April. And which I decided I would go to. And a couple of weeks later when the day arrived: off I went, enthusiastically, and curious' about what I would further learn or find out.

The conference itself was an all day event, starting at 9.15 a.m. and finishing at four in the afternoon. And which the main theme of was how to understand and cope with 'hearing voices. And was for the purpose and benefit of both, people suffering from this problem, and, professionals associated with the mental health field: such as, mental health nurses, psychiatrists, psychologists, social workers, counsellors or affiliated professionals. 'Or organizations who wished to set up a voice hearers group. And it comprised of several talks at different times throughout the day; and which some of took place in a main hall. And were conducted by both: 'people who had done a lot of work in this field, research etc. And also some people, who were what I would refer to as victims – who had clearly been invaded by one of these Entities. And who,' were giving their accounts and history of their own suffering.' And additionally to this, there were other types of lectures and discussions being held in other rooms throughout the building, throughout the day, two of which I had my name down for, and attended. One called: Coping strategies,'' and the other: The writing on the wall'.

But the fist talk of the day was in the main hall. And as well as it being an introduction, it also went through a brief history of how long this problem had been around: and mentioning some well known historical figures, who too, had lived with and been troubled by this problem. This itself was interesting! But towards the end of this initial introduction and history of the voice hearing experience, something else was said that had really stood out. The final speaker had said: "That it is not known exactly what causes the voices!" And to me, as I sat there in this hall, with some five hundred or more people; 'and many from the medical professions, to other such people, who would think themselves as experts,' or, with a high degree of knowledge and experience on this subject; I was the only one, who knew the answer to this statement and the absolute truth; in all its detail. "And, this only strengthened my resolve!" Further inspiring me to start work on the book; and reveal, the true reality that lay behind it all. "And everything now seemed in place to start the writing." And on the 18th of April just over a week later; I sat down at my new table in the kitchen; and began to write my story. And which I would title the name given to me by the Entity: "The Unwanted Companion!" And I would start where it all began; at the beginning, the day I came into the world. Allowing the reader to see that I too had had a life, and aspirations', just like other people. But also to see, that whatever path we plan, or choose for ourselves in this life, all may not be as it seems. And that during, the twists and turns of life we can expect the unexpected; and may well end up on another path; and not always one, that is desirable, or of our choosing.

And although I had started the writing; it didn't by no means mean that I trusted that the Entity would not cause problems for me! And was always fearful that at some point it would turn nasty' and either rip up my work during the night or do it, right in front of me, like before. And so I planned, that after every ten pages I completed I would photocopy them and give a copy to Margaret Palmer for safe keeping, just to be sure. 'And the first ten pages were completed about five weeks later.' And although it was slow moving, I was quite glad that I had finished these first pages. And it was also harder than I thought it would be,' and required a lot of reflection, and thinking back to those early years of my life. But I was right about one thing and that was that the Entity had started to help me – giving me help in small doses. But there were also times when it would hamper me as well. 'But overall, the situation seemed more positive than negative. And one thing I was pleased about more than anything; was that I had started the book, and it was getting written.

And besides this more positive phase of developments and the getting of the book underway. There were other aspects of my life I had been trying to improve also: "Partly as a distraction from my unusual situation." And part of which involved me trying to participate in life once more after the many years of misery that I had been put through, and which had stopped me from doing so. And so, I made a list of things I wanted to do, or places I wanted to visit in London; and knowing that at least I wasn't stifled anymore by the physical health problems, and could now do them. And one of the first things on the list was to see the longest running play in Britain. A murder mystery called the 'Mousetrap' and which at this time was fifty years old and was on in the West End, and which I went to see. And another thing I had done later during the summer was to go and see Shirley Bassey in concert at the Wembley arena. And which was superb. Hearing those songs live with an orchestra added another dimension to them. It was brilliant night. And I also went to see the actual making of 'Top of The Pops' on two occasions and which was also' another enjoyable experience.

"But also during the beginning of this summer, and partly for two reasons, and one of which was experimental." 'I had started taking a drug called Clozaril.' Which was more relatively new on the market; and which was prescribed for schizophrenia! And prescribed only after the patient had first tried at least two other anti-psychotic medications.

I had heard a lot about this drug, being told it had good results in those that heard voices. But knowing what I knew was baffled at hearing this. And my main motivation for wanting to try it, was that I thought that maybe it affected something in the body, that may of dampened down the ability of these 'Entities' to operate in their usual way; "and hoping that this was the case wanted to see if this was true?"

And although the Entity had already told me it would have no effect: "even telling me, to go ahead and try it and find out. And which I decided to do just that, and try it anyway; and also knowing from what I had been told that I would have to wait at least six months, before I could see the full effect of this medication: 'needing initially to gradually build up to a certain dose! "And for me - also see if it would have any effect on the Entity." And at this point I had been taking it for two months; and so far as of yet it had had no effect on the Entity, or its abilities.

But apart from the progress I seemed to be making in some areas, there were still parts of my life that had not changed. And one of these; was that I was still seeing my social worker for an hour once a week, almost every week. And, although, I so much just wanted to move on from this psychiatric world, that had become a big part of my life. I often found, that it was also, the only part of my life, where there was some constant support and help available; and where I didn't feel out of place. And, at this particular time, it was this weekly visit to see my social worker,' that gave me somewhere to go; and during the worst times, kept me sane. And, as well as these also occasionally being home visits, we would occasionally go out' and visit places too; for an afternoon etc. And, it was my social worker and the Psychiatrist' who had first informed me about the Clozaril, and arranged for me to try it.

And although I had known that at this stage, if I again attempted to tell them the whole truth; "I was almost certain,' that once again I would not be taken seriously? And so, when asked how things were during my weekly meetings with my social worker, or during the three times a year I saw the Psychiatrist, I told them the truth about what had been going on and how I was feeling at that particular time. But not saying' that I knew for sure, that it was all the result of being possessed by an 'Evil Spirit Entity'. And instead, just mentioned the symptoms, and problems that it would cause and how they had been affecting me. And if during those times it had been particularly bad, would explain things in a way that generally left out my having to mention the Entity. So, all in all, apart from the fundamental reality behind what I would describe to them as symptoms; I was telling them the truth; but, "being selective in what I told them!" 'And, the only difference between me, and all the others in the same situation, was that I knew the truth; and the real source of what they thought was mental illness.

'And as for my cured physical health problems:' when my social worker had often enquired on their progress and how they were. One day asking how my legs now were. I had said they were fine now and had gradually got better: 'to the point where I no longer needed to use the strapping on my knees.' And as for my back, I had said that I was continuing to do both, back exercises, and sometimes go swimming, and which,' had continued to greatly improve it. And that it too had vastly improved! And as for my eyes, when asked at a later stage' I had said that I was still doing the Bates Method eye exercises; and which were having a good effect on them, "saying that they too had vastly improved and now I was able to read things much better and do other things. And, whether it was my back, or eyes or ears I was asked about, I now only ever talked about these former physical health problems' when I was asked about them; and their condition, or progress; and did not mention them, unless I was asked! And kept what I said short! "And although I hated having to lie or be evasive, in reality, at this stage, what else could I do? After all, I had already tried to tell them that I was possessed by an evil spirit entity and they hadn't taken me seriously. And had looked on it through what they perceived was the illness talking. So what could I say...? "Oh and by the way", that evil Entity I told you about: it had cured my health problems as well! What kind of response would that have got! And as I said, that the only difference between me and all those other poor bastards who had been invaded by one of these Entities, was that I knew it, and knew the truth about it. The others, were in the dark and would be kept in the dark, and would never know. And so, I remained silent about what I knew about my reality, and the others that were unfortunate enough to be in the same situation. Knowing that at this stage, it would not have done me any good, or benefitted me in any way,' to have said anything else.

But although I steered clear of saying the absolute truth, and knowing also that I wasn't mentally ill, I no longer referred to it as 'schizophrenia' either. But had at times, as time had moved on, on occasions, said that I thought that it was a spirit of some kind that was causing all the problems, and instead, often referred to my situation, as "the problem". And it wasn't just for 'my benefit that I had a weekly visit with a social worker or still had to see a psychiatrist three times a year. I was under the 'care in the community programme.' And this was also to ensure, that you were either: 'taking your medication', or, not a danger to the general

201

public, or yourself. With a diagnosis of schizophrenia they tend to keep an eye on you.

But I did plan to tell them the whole truth. And this would be when I had finally got rid of the Entity. And which I would then be able to walk away from the psychiatric world. But until that time arose, I felt at this point it was not worth trying again to convince them otherwise: of the reality! As there was absolutely nothing they could do about it anyway. And knowing what I knew I had also said to them that I knew my problem was not going to be cured by any type of medication; 'and that I could not live with it either.' And that I intended to go back to India, to see Sai Baba again, to ask him to get rid of this problem for me. But had also said that there were tasks I was almost certain I had to do before I could be released from this burden, and could not go back to India until I had done them.!"But did not tell them what they were!" And just referred to the tasks, as being of a spiritual nature, and that they were personal. 'And what they thought of that I don't know'.

And these tasks I refer to, were, "the writing of the book." And which was by now, part of my overall plan that I had now put together. Coming to the conclusion that I would first write the book, 'and then' go back and see Sai Baba. 'Evermore increasingly feeling that maybe this is what I had to do.' And, overwhelmingly' feeling that it wasn't worth going back there until I had done this.

But, apart from this whole scenario I was dealing with and involved in, I was also very grateful to the care in the community services. They were people with integrity, and who generally had my best interests at heart. And, seeing a social worker once a week was generally a positive thing; and could be helpful in many ways. And at this stage was still a much needed crutch.'

The months were going by, and it was now October. And although progress on the book was slow moving: 'having only completed 28 pages'. It was nevertheless, often enjoyable, and somewhat strange to be going over my life; writing about it; and all the memories of things that had resurfaced and it had brought up; many of them fond memories. And which often made me stop and think more about them; reflecting and reminiscing, and enjoying the moment, through what seemed like a window to the years gone by.

And although some months earlier when I started the writing, I had thought that I could probably finish it in eighteen months: 'at this rate that it was getting done', it looked like it would take a lot longer! And I now realized' that if it was to get completed closer to my estimated time: I would have to put a lot more time into it.' But at the same time, I didn't want this to dominate my whole life, and badly needed some other outlets, that would give me a break from all this. And also help me to participate back into some areas of the normal world. And one thing I had done a month earlier in September, to help this process, was to enrol in a language class for one evening a week, learning German; and which I had already had some knowledge of. And I had also been thinking of getting back into the dating scene; finding a girlfriend!

But it was also around this time, later towards the end of October, that the Entity had started showing me more, or other aspects of its abilities. And using them in a nasty way to either wind me up or just generally be evil. It had been altering the positions of some of my teeth; one day showing me what it had done to them while I had been looking in a mirror. And from what I had gathered this had been done during the night while I was asleep. And it had also been making my hair fall out, and also showed me where it had done this too, and showed me where the hair was missing. And it also started to do the same thing to my beard. And I don't mean a full grown beard as I was used to shaving almost every day, but I started to notice that this too was being tampered with; like a small section' or patch of it had disappeared! It pointing this out too while I was looking in the mirror, and confirming it had done this by making another small patch disappear, so I could see it being done right in front of my own eyes, 'the hairs disappearing, leaving a bare patch. And as my beard is black, it was very noticeable, and made a noticeable difference to its shape and appearance. And these things not only made me feel very angry, but were also distressing. 'And although I had got use to', and put up with many of its aggravations! It was evil stuff like this, as well as the other annoyances of differing extremes, that made me question, if by writing the book first: 'was I doing the right thing? And at times, made me feel like just jumping on a plane to India, and staying there', until I got to see Sai Baba.

And as time had moved on, the Entity had also started using its "delusional abilities" more, to further annoy me or make my life generally more uncomfortable. Tricking my thoughts, and perceptions, as well as causing paranoia, which it could turn on and off like a switch. And which would sometimes affect me to the degree that I wouldn't go out until it got dark; as in the daylight I was being made to feel like I was in the spotlight and under scrutiny by everyone, and felt extremely self conscious. And I sometimes wondered if these were done not just to annoy me or make my life difficult, but also to further demonstrate how it could affect my life in all manner of ways? And examples of this would present themselves in different situations and often causing me to feel anger or aggressive towards other people: often being made to feel like I was being watched, or laughed at, or spied on, or followed, 'or people in cafes or restaurants were talking about me, to the point where I would often turn around and stare at them or stare them out; and not even caring if it resulted in a fight.' So worked up, and consumed by the moment of delusion, that I was willing to go all the way, backing up the delusion with violence.

"And during these times", I had really been convinced that I was right about what I perceived; 'sometimes feeling murderously violent!' But later on when I had removed myself from the situation, or had come home: and after gradually realizing that the Entity had been the cause of these incidents.' On realizing this,' it would often sarcastically say: "sorry about that!"

And although I had gradually became wise to its games of this nature and did my best not to fall for them; such is the power of these Entities, to consume, and control, and delude, that most of the time, I couldn't do nothing about it.

But one thing I had developed: if I thought I was being worked into a situation, I would leave where I happened to be. Or another thing I would do to fight its influence was to avoid making eye contact with people. But these things were often to no avail, as it could soon distract me and cause me to look at someone or something when I wasn't paying attention, or had relaxed my guard, it trying to cause trouble in this way too. And living like this, often seemed like hard work, and too much of a burden; as it often seemed like there was no peace in my life. And I had even become concerned that if this kind of thing carries on, then at some point I may end up in a fight, or attacking someone, and end up being locked up – 'all because of the Entity'.

"It often seemed hard to fathom how it did what it did." And equally mind boggling was its knowledge and workings of the anatomy. But this is what these Entities' can do. They can totally delude your thinking and perceptions and get you in all kinds of trouble. And although to some, it may even sound amusing, it is however no laughing matter. And like I've already said, these delusional abilities can be switched on and off and used in different strengths, and can easily be used to make a person attack or kill someone. And these' were just a few of the ways that my mind was now being played around with, and had been for years before – 'without me even knowing about it;' but now even more severely, and in other ways, and other situations; and maybe to demonstrate and further reveal to me, how easily these Entities can take control of a person. 'Maybe further confirming to me what it had told me in the revelations! "Although I didn't know for sure if this was the reason, and it could quite possibly have been for its own amusement. But I was realizing more and more just how quickly this delusional process can be put into action. And to put it simply, these 'Entities know what buttons to press'. And when they want to control' or delude the person, 'they merely press them.'

And although the care in the community service,' a couple of years earlier had given me a brief introduction, to one of their psychologists, who I had had two meetings with. It was also during and around this same time, because of all this, and how my growing concern about it, had often been expressed to my social worker about what had been happening, that appointments had been made' for me to see another psychologist.' Them,' hoping this could help me in some way! 'And which I felt obliged to go to for a while.' But also now feeling it might in some way help me'. And which although some of the overall thing was beneficial, I knew also I couldn't tell the whole truth, and therefore, ultimately knew at best, it was cosmetic, and only a diversion.

But it wasn't all bad, or like this all of the time. And there were periods of time sometimes lasting weeks on end, when the Entity would not really bother me in this way or subject me to these things. And I somewhat consoled myself by thinking that what it had done to my teeth, hair, and beard; "that it might eventually put them right again, just like it did with the health problems; and, hoping this was the case". And despite these worse times, I was still eager to carry on and get the book written. And as the months went by, I was now becoming more aware, that the help I received from the Entity during the times I was working on the book' was increasing; and excellent. And the often brilliant displays it could produce in terms of help, ranged from: its brilliant memory of things, and events in my life, to helping me move the story along. And despite all the troubles along the way, the months were going by and the book was getting written. And a new year had now dawned. It was now 2004. And in February,' I had decided that I had now given the Clozaril the full test; "now close to a year since I had started it." And had fully realized,' that this new drug that was spoken so highly of; would not, and did not, have any effect, on the abilities of the Entity whatsoever; and around this time, had stopped taking it. 'And I had only stayed on it for this long', because, during some of the worse times, 'I had hoped,' that it would have an effect on the Entity's capabilities.' And I think that the Entity had not only encouraged me to try it, but to also make the point and further enlighten me to the reality of what it had already told me: 'That medication would not do anything to affect the abilities of these Entities!" And like it had told me previously, that any moderation in the symptoms, or improvement in those that were possessed by this type of Entity, was due to the 'Possessing Entity moderating its behaviour." In order to give the impression that the medication is having some effect. And usually doing this to keep all those involved ignorant and in the dark.

But even with this moderating of their behaviour, they will still subtly manipulate, and still play around with their victims mind in all the usual ways. But producing the 'symptoms in minor ways, giving the impression,' that things have improved because of the medication.' But sporadically from time to time, will be very nasty and produce the more aggressive and worst forms, of the paranoia, delusion, the voices, the trickery, the insomnia, hallucinations etc; all the usual stuff; in a sense making the situation unpredictable! And sometimes even influencing them to stop or forget to take their medication, so it seems like this was the reason for it. And as I had gradually come to learn: that for those who are unfortunate enough to have fallen victim to one of these Entities,' that long term, the only thing that is predictable', is the unpredictable! And this can either be on the minor end of the scale, or worst end of it, 'all depending on the personality of the Entity'. In general, the best thing some of these medications will do for these people: is that the often somewhat tranquilizing, or numbing effect of them, will reduce some of the anxiety and distress, felt by the individual.

"And although I had now stopped taking the Clozaril." I had not informed my social worker,' or the Psychiatrist that I had stopped taking it. And one of the main reasons for this' had been because of a 'growing number,' of high profile, random murders being reported in the media. By what was usually described as mentally ill people; and often involving the use of a knife, or some other weapon. And, usually stating the person had a history of mental illness, and had been diagnosed as suffering from schizophrenia; and, or voices in their head had told or had commanded them to do it. 'And additionally reported, that some of them had stopped taking their medication'.And which, had often made the national news, or were reported in the newspapers. And in a three year period there were at least a dozen of these murders that I had heard or read about myself. And had started to cut these out of newspapers and put in a scrapbook. And because of these, and the growing public pressure, new laws were now being drawn up to be put before a vote in parliament. And which would confine and detain those who were known to be what was termed mentally ill – and had refused to take their medication – or medication in general. So with this in mind, at this stage; it was much safer for me to convey the impression, that I was still taking it.

But the good thing that had come out of my trying the Clozaril,' was that I had got to talk to many other people who were also taking it. 'And not one of them', had said to me, that it had got rid of the voices!' And these were mostly people I had got talking to' while I sat in the waiting room, for the monthly blood test at the Clozaril clinic'. And which is a part of the monitoring process that comes with being on this medication. 'Due to a side effect that can occur within the blood. And which at first, the blood test was every week: then after three months, eventually becoming once a month. And it was mostly these people who had spoken to me first, and usually asking questions to do with the medication – was it helping etc. And what I gradually began to learn was that most of the time they were referring to the voices: if it had stopped them or had had any effect on them at-all. And as time passed, sometimes I would talk to and ask people myself. Ask them what their symptoms were etc. And if they mentioned the voices I would ask if the medication had got rid of them or improved things. And the usual responses were that it had either helped initially, or for a while,' it seemed to have dampened them down. Or that they were not as bad, or they didn't appear as often, or that they would still come and go. 'But never:' got rid of them!

As was said, some of the Entities' will play along with the medication, and moderate their behaviour accordingly.

And as I was still attending the hearing voices group, I also got to know other people who were also taking Clozaril, and their answers were no different, with some saying it had had no effect at all.

But during the times while I had sat in the waiting room, waiting my turn for the blood tests, I had also talked to people on Clozaril who I could clearly identify as 'not' being possessed by one of these Entities', but were suffering from a brain disorder of some sort. And it was some of these that had told me that the Clozaril had dramatically improved their lives and that they had really benefitted from this drug. And although this was good for them, the trouble with the label schizophrenia,' is that it encompasses a wide range of symptoms that fall under the same banner! And usually the only distinction that is made,' is the difference between diagnosing someone to be suffering from 'schizophrenia, or a paranoid schizophrenic?' And so there needs to be a new diagnosis incorporated into the process of diagnosing a patient. And that is one that I have termed: 'Possession Schizophrenia!' "For those, that have been invaded by one of these Entities." And the reason I link the two names "possession and schizophrenia", is because the trouble caused by these Entities is always of a delusionary nature; and because of this these people usually end up in psychiatric institutions or hospitals, or affiliated organizations. And for this reason it needs to be known, exactly what category this person is in. 'And the diagnosis' the person fits into. And although there is a clear group of symptoms' that clearly identifies and distinguishes the type who is possessed, 'with the main indicator being the voices. Unfortunately for the time being there is no concrete way that I know of that would be able to remove these evil Entities from their victims, and to also prove conclusively that they have been removed and have gone. But this area', and the recognisable group of symptoms, for a specific diagnosis, is something I will talk more about later; as well as in terms of what could be developed to combat them.

But it was my accumulated knowledge I now had on this subject, not only through what I had been put through myself. But was still often,' being given more bits of information by the Entity; further elaborating, demonstrating, emphasizing, and pointing out things, in order for me to build a clear picture' and profile, of the 'type' that is possessed by one of these Entities. And even, if they had not yet experienced the voices.

Living with this Entity? 'It was a strange relationship that it had become! And as I mentioned earlier, was also an evolving one. And this had also further evolved in the area of communication. And the methods of communication', had by now, become a finer tuned blend of all the components. And of which' had now become a well established pattern, that I had now become fully use to. And a further development of this was the comments or things,' now said to me through my own voice. And this as I had talked about earlier had initially started some time back. And initially' what I had first been tricked into thinking were Angels. And during that period, had usually been a yes or a no comment. But as time had gone on it had now also become another way that the Entity would now use, in various ways, to make comments. And which were generally short statements, and could be about anything. And although it sounds eerie or weird, and which it initially was and which I hated, because I now knew it was the Entity. But it was also something I had now become use to too, and no big deal anymore. And the only time it now made me feel uncomfortable, or felt eerie, was when it was done during anytime I happened to be looking in a mirror. Because during this process, my body would be taken over, and it was just too much to be looking at myself, suddenly being taken control of, and used and animated, and it not being me that was doing it. And this was something I never got used to. It was too bizarre. And whenever this was done in front of a mirror I would do my best not to look, or try and turn away. And, the Entity knew I didn't like this at all, and would also sometimes use it to annoy me or make me feel uncomfortable.

But a typical conversation with the Entity, or delivery of information from it would usually start by it first producing (inserting) the thoughts into my mind. And this was the most used of the methods. And whenever it occurred' the slight energy that came with it was instantly recognisable. And would either be used to make a comment or suggestion, or tell me something. And depending on what had been said to me or what I had thought, or if I'd asked back in response, elaborations would often follow this: by either inserting more information into my mind in the same way, or would often be blended with images -'being flashed into the minds-eye,' or the voice in the head, or the sometimes talking through my own voice. Generally, whichever was the best way to say or explain something, or get a point across, or the most convenient; depending on the situation, or where I happened to be?

Or when it was making a point of some importance. 'Whether this was a substantial insight into something, or a specific valuable piece of advice. It would now use its ever present instant ability to take control over my body, and after making its point, would slightly nod my head back and forth in a slight rocking motion; 'and which the presence and personality of the Entity came through with it. And while at the same time as the nodding, further elaborations of information, on its point or advice or observation were flashed into my mind, either in images, or the thoughts fed into it; further cementing that what it had said was sound knowledge, a good idea, or good advice, or the best option. And the nodding would usually continue until I had recognised and acknowledged the importance of its advice or insight. And because it knew all my thoughts and everything I was thinking, it was generally done without me having to ask any questions, (but obviously sometimes I did).

And this was something that had now become a familiar part of the whole process of communication, and what I was use to. It had become a fine tuned blend of all these things. And another more humorous attribute of this was: sometimes when it had made a quick point, or clever observation about something, it would just simply wink, or just do a single nod.

"And although I was living with the bizarre and the highly unusual; it was this bizarre and highly unusual that had now become the usual and the normal, and just an everyday part of my life. And it no longer seemed as strange and bizarre as it once was." It was something I had now got use to! And although there were the ups and downs with this situation, there were also often times when the Entity was good company and a good source of information, often making wise contributions to my thoughts on various subjects, 'or observations, or world events; often displaying another way of looking at something. As well as giving me tips about human nature and psychology, and people in general, and general good advice, it displaying exceptional levels of intelligence, wisdom, and insight. And this could also manifest in practical ways too, some furniture I had been making for my flat, it had helped me with, by giving me practical design ideas, producing images of different designs, and giving me further tips during the designing and building process as well.

And another part of this more positive side of it was its sense of humour: And which it often displayed in these same capacities'. And which could be anything from amusing to hilarious. And in the same way it could spot flaws and make sharp observations about things or people, and had a highly developed creative intelligence. These same attributes often came through in its humour to. "And it was this, and the blend of the more positive and the often helpful side of its character, that had often made me aware that the Entity' could also be good company.

And the humour of these Entity's, and mischievous nature of it, was also something I had heard expressed by others who were unknowingly possessed by them; either in the hearing voices group or accounts I had read from other people in the form of information I had come across. And the type of thing I had heard said was that: 'Sometimes the voices say things that are funny, and they make me laugh'. And similarly they also said, that as well the voices saying negative or nasty things, they would also say wise or helpful things.

But these Entities,' will also convey humour using their subtle method of inserting thoughts and information into the mind, without the person being aware of it. And which the person will suddenly find themselves bursting out laughing because of this. And this was something that was often done to me during my years in the dark: and which was later pointed out to me by the Entity. And one woman who was talking about her symptoms at the hearing voices conference had said exactly the same thing, 'that she often found herself laughing, sometimes hysterically,' and had said that when this would now happen her mother would say: 'oh, she's off on one again! 'Meaning they were use to it. "And as more often than not, the humour will be of a crude or mischievous nature, and if asked what was so funny, 'you would have to be economical with the truth.' And similarly, and in the same way as this aspect will sometimes manifest; the influence of these possessing Entities, subtly filtering through their victim can also be seen in signs of high intelligence and creativity; this influence manifesting itself and coming through also.

And; as I was saying about how the situation had further evolved and how the Entity could also be good company and how a typical conversation unfolded. And although the general scenario would be that it was usually the Entity who initiated comments or suggestions or conversations. There were also times when I would be thinking about something; a situation, a problem etc, and as a last resort, and often begrudgingly, I would ask the Entity for advice, 'saying for instance: 'what do you think I should do', or what do you reckon – any suggestions? And what I often found during these times was that it would soon produce two or three ideas or options of things to do. And then run through each one in the context of the problem; and, finally pinpointing the best one or best course of action; 'and further elaborating on the reasons why.' And, although it was an unusual situation I lived with. It was during these times that I was often reminded just how unique and special and useful this all was too.

"But these were the better times," and as you can see, life with the Entity was a mixed bag. And often, even one day could be a mixed bag.

Chapter Ten

The Noble Task: Part Two

And Template for a Diagnosis and

Combating Them

The year was progressing along. It was now spring 2004: And now a year since I had begun to write the book. 'And as far as the book was concerned,' I was still making slow', but steady progress. And as I mentioned earlier I was also trying to have other outlets that would divert some time and attention away from the book and life with the Entity. And it was around this time that I had been thinking more about finding a girlfriend; and had decided that I would start going to bars or other places where I was most likely to meet someone. "I certainly wasn't desperate to meet someone, but was more or less testing the water once again." And I had learned and noticed that when I was with people that most of the time the Entity hardly ever bothered me, and I often forgot about its presence in my life. So it was also about bringing some more people or normal situations back into my life: trying to counterbalance what I was having to live with; and also improving the quality of my life.

But my aspirations to try and meet a girlfriend were short-lived. And this had been the result of the Entity ruining my chances. And, had first set about doing this, by distorting and changing the way my face looked!

I was already aware of its capacity to do this, as it had demonstrated it to me before, showing me just exactly what it could do. 'But now it was doing this, and saying', that if you try and get a girlfriend, I will ruin it for you: I will change your face so you don't look attractive; and change it in a way that nobody would want to speak to you: First saying this to me one day while I was looking in the bathroom mirror and combing my hair; 'it changing my face so I could see it.' Then telling me this is what it would do. And further saying, it had often changed the way my face looked when I was out and about. 'And although, it had already demonstrated its ability to do this before, it was now: giving an even more varied demonstration! 'Which included' altering the shape and bone structure of my face; and shape of my jaw, chin, and nose: As well as producing the more weird, or bloated; or nasty looking expressions' it had already shown me." And even produced odd and unflattering looking lines on my face, and just as quickly made them disappear. And what really stood out was the speed at which it could do this; "literally in the space of a few seconds;" it could change the look of my face and almost just as quickly change it back again. 'And what started to soon happen was that it would continually be demonstrating this almost every time I had to use a mirror.' And then also started to do this in other places besides my flat', 'showing me how it had changed the way I looked,' whenever there was a chance to see my reflection in something.' And the gradual effect of knowing how fast this could be done, and that it would do this anywhere – often making me look like a person I didn't recognize. The effect of this; 'was that eventually I lost my confidence'.

"When the Entity had first said it was going to do this to me, I had initially tried to brush it off, and not let it get to me or bother me. Feeling that it was either just threatening this, or trying to control me, or that it was just annoying me: like it often did? And I had gone out on a couple of occasions with the intension of trying to meet someone. And on the second occasion,' it had not only changed my face, but had also caused me to trip over, 'making me fall on my hands and knees, in a bar, in full view of lots of people! And which after getting up: I didn't bother to look around at the reaction, but just left. 'And thinking what an evil bastard this Entity was."

'It seemed that it was saying: "that not only will I change the way your face looks," but I will make a show of you as well?' And being tripped over in this way was nothing new. It had done this to me twice before, a year or so earlier: once making me fly into the back of someone in Oxford Street. 'And another time, after crossing a road', causing me to trip on the curb.

But after this incident, I had decided; that as long as I had the Entity with me, 'that looking for a girlfriend', would not be possible. "And that was that!"

But besides its ability to change my face or tamper with any part of my body, a further demonstration of this ability during this time had incorporated changing the way my eyes looked, and changing them back again. And it could also make 'them' look different in all manner of ways too. 'The worst of which' was giving them an evil, nasty, or abnormal look about them. And my first noticing this after getting up one morning and looking in the bathroom mirror, and which I was soon made aware of what it had done to them: noticing the difference straight away. And which soon after, 'I had then also been reminded; and prompted to think back to the time of the encounter with that couple at the bus stop: during the time' I was being made to hold the cross, in the palm of my hand.' 'And, the unnaturally evil looking eyes of the man who had spoken to me; 'and by what he had told me' I had realized that he too was possessed' but didn't know it.' And it was now also, while I stood there observing my own eyes, and just how effortlessly they could be changed, that the Entity had told me; "that his eyes", had been made to look like that, by the possessing Entity! 'And his eyes' did look like pure evil.'

So as well as the other trouble these Entities cause to those they possess; they can also change the way the person looks: and usually for the worse. And so not only had it been making me look unattractive, or weird, but was sometimes making me look nasty with it too. And this itself was causing a complexion; as, unless there was a mirror right in front of me, I didn't really know how I looked at any one time. 'And by the strange looks I had sometimes got from people', it made me wonder what it had done to my face.

But apart from the giving up of the idea to meet someone; there were other adjustments to parts of my life. And by now, I had stopped going to the hearing voices group as often as I did. And by the summer of 2004, was only showing up there probably once a month, if that. And, 'the main reason being,' that I didn't need to be there as much as I once had.' And it was also during the later part of the summer that I had also stopped going to the Sai Baba Centre too. 'And one reason for this;' was that the Entity had increasingly been making my time there difficult.' And another reason was that I now wanted to spend more time on the book; and just concentrate on getting it done. Knowing that every day that I could work on it was valuable. And I had decided' that getting the book finished was more of a priority, and that: I would now pray at home instead. And I had a good supply of books on Sai Baba' which I had bought while in India; and which I was in the process of reading. And, a couple of months before I had stopped going to the Centre, I had also subscribed to a monthly magazine: to do with Sai Baba and his teachings; and which was produced by another Sai Centre. "So, Sai Baba was still very much a part of my life!" 'And, after completing the first year of the language course', had decided, for similar reasons', not to do the second year; also realizing this, and the level of homework involved, was now, 'also taking up to much valuable time.'

"But it was also something else as well which had given me the confidence to stop going to the Sai Baba Centre and think that it was ok to do this." And this was an extraordinary set of coincidences, which had first started during the previous summer of 2003. And at first had me baffled as to their cause, or their origin, or even there meaning: the first one occurring in August. And which I later discovered, "were a calling card used by Sai Baba". 'Often used to demonstrate his presence and awareness in the life of a devotee: or someone who was becoming one. 'And the first of these coincidences,' had occurred one day after coming out of a department store, in Oxford Street; then soon after getting on a Double Decker Bus! And, while on the top deck could hear a conversation going on between three American women. "One of them was trying to remember the name of a place in the American state of New England." And which,' had been notorious for the Witch Trials that went on there in the late 17th century!

But it had only been about 20 minutes earlier,' that I had come out of this department store. 'After having been browsing through the books in the travel section', and in which I had picked up a book about Boston; had looked through it, and had stopped on a page, that spoke about the Witch Trials in Salem, 'New England' and which I had read a bit of. "And not long after I had left the store; and then getting on this bus and hearing these women saying (quote) 'what was that place in New England called, where they had the Witch Trials? And although none of them could think of or remember the name, I knew it was Salem,' and had just been reading about it. And although I felt like turning around and saying, "it was Salem," I didn't. And that wasn't the point. The point was the coincidence and weirdness of it at the time it occurred. 'And although it had felt strange,' I had also later thought that it was just a very odd coincidence? But this was just the start! And by the others that followed; I soon began to realize, that these were more' than just a coincidence?

The next one happened only a month later, on the day after my birthday. I had gone to see a play in the West End, called 'Hitchcock Blonde'. And after having purchased the ticket at the theatre, and having about an hour to kill before the start of the show 'had decided to go somewhere for a cappuccino. And had walked down to Borders the Bookshop in Charing Cross road, which was close by, and where I knew they had a coffee shop on the second floor. But as I walked into the main entrance of the shop, and past a row of books, 'one of them had caught my attention,' and I had stopped and had had a look through it. It was the autobiography of a well known British actress called Amanda Barrie. I had remembered a specific film that she had been in during the 1960s called 'Carry on Cleo'. 'And, I had browsed through it,' looking at some of the pictures of her, during the early stages of her career? And one picture I came to, had said: "That this was the first time that I saw my name up in lights." And it was during a time she had been performing at a theatre in the West End. And when I looked at the picture: "It was the theatre that I had just bought my ticket from and was going to see the play in; the Lyric Theatre in Shaftsbury Avenue." 'And which' I had just come from five minutes earlier.' And, although, this second one', had also seemed a bit more than just a coincidence; "I still didn't know what to make of it?"

The next one followed only a couple of weeks later: occurring one night while in my flat. And not long after I had watched a video, which I had bought some months earlier' called, 'The Stones in The Park'. And which was of a free concert, performed by the Rolling Stones in London's Hyde Park in 1969. 'And during which, the lead singer, 'Mick Jagger had read out a poem by a poet named 'Shelly.' And which, he had devoted to a member of the band who had died a couple of days earlier! And before he had read this poem out, he had first named the writer of the poem, as being: by the 'English Poet, Shelly.

But later that evening not long after I had finished watching the video; I had turned the television back on to normal T.V. 'but at the same time wasn't particularly watching it.' But then I decided to switch the channel. And the programme I had switched to, 'had been about the comedian, Billy Connelly, touring around Britain on a motor-trike, 'and stopping off at places of interest along the way. However, the moment I switched over he was standing in an 'old grave yard somewhere'. And then a moment later he said: That this was the grave of the 'Poet "Shelly". 'And what startled me about this, was that I had bought the Rolling Stones video months earlier; and had chose to watch it for the first time that evening; but could of picked any evening to have watched it; or even any moment to have changed the television channel; and, if I would have changed it a few seconds later; 'I would have missed that? "And, it was after this;' that it started to seem, that these 'coincidences', were just too much of a coincidence, "to be a coincidence."

And then there was more of them, and which occurred in similar ways; sporadically presenting themselves every so often; and now' occurring more often. And some of which had had quite a startling effect on me: by now feeling and realizing by their unusualness, and number, that they must be being orchestrated 'specifically' for me. And then for a period of about two and a half months there hadn't been any. But then the most startling one of all occurred! It was one evening during the late spring of 2004, and almost a year after they had first started, that on this particular night I had been working on the book and was busy writing; and had written something that sounded familiar, and had wondered where I had heard it before. And in the writing I was explaining how I had come outside of this building: *it being during the time of the voyage through hell; and on the day I had had my eyes examined in Harley Street.*

And had written that after leaving the building,' I had come out into the street and the sun shone'. And it was this ,"and the sun shone" I had pondered over; from where had I heard it before. And eventually decided to change 'and the sun shone' to: the sun was shining! But at the same time' kept a note of it, thinking that I may later change it back. And although I had put an unusual amount of thought into this, than I normally would have, over just a few words, I then carried on, and thought no more about it. But it was the following night, and the last night of the language class, after I had left the collage and got a bus to the Islington Angel, and had gone into a Borders Book Store there: just to use the toilet. 'And as I had walked towards the toilets; situated right next to them, was the section of books, to do with crime and true crime. And as I got closer, had noticed a book 'written about British gangsters of the last century.' And for some reason it caught my attention; and I thought I will have a look at that when I come out. And when I did so, I picked up the book and had a browse through it. I then stopped at a certain page; looking at a photo of one of these gangsters. And, as I had looked' at what the writing had said under the photo: and, it had been a continuation from what was being said on the previous page.' And the rest of the sentence, starting right under this photo, said: "And the sun shone!" "Exactly' what I had pondered over the night before." And there it was, staring me right in the face; 'and the sun shone'. And when I actually seen this, it literally sent a shiver up my spine, 'literally', and stood staring at it in disbelief. I then closed the book quickly and left. It did actually freak me out.' And this wasn't just because it was another coincidence; 'or the complex mechanics' behind it all. But because it was like a stamping of something; and that it seemed to close for comfort; and confirming and really bringing home to me that there was something at work in my life besides the Entity, and it was watching me and guiding me to these things; and that, it wasn't just the Entity that I was living with; but obviously something else was at work too. And, as I eventually got to learn; that coincidences,' and more specifically, coincidences like these, "were one of Sai Baba's calling cards." And in the way he does things, these coincidences were measured, and the content of them becoming closer and closer, gradually building and producing them to the point, where "bang" this latest one had stunned me and had made me realize in no uncertain terms, that they were no coincidence! And it was this one, which had had the biggest impact, and brought home this full reality. And I thought about this, on the bus journey home; and even after I had arrived home, realizing more than 'ever', that there was a lot more to life than we ever realize. 'And bringing home this full reality' of something that I already knew, and was aware of for obvious reasons'. And that is how we can all be guided and manipulated in this same way, without are ever knowing about it; 'thinking all along that it was ourselves that had made all the decisions'. And, 'how we can also,' be manipulated and guided into all kinds of situations, for good or bad reasons? And when you hear people talk about fate, i.e. 'we were meant to meet and be together'. 'Or they just happened to be in the wrong place at the wrong time.' It really brings home, how an unseen guiding force of fate could be at work; and bring this all about!

But this wasn't the end of the coincidences, and they continued. But not as often; and probably because I had got the message! And, had concluded, and recognized through various additional means; mainly similar stories I had heard and read about them also happening to Sai Baba's legions of devotees: 'That they were the work of Sai Baba!' And that the workings of coincidences were one of his calling cards. Letting know or reassuring the person, that he is aware of them and their problems or situation etc. 'Or simply just proving or making a point! As some coincidences,' could have a theme, directly relevant to things they had prayed about.

And of the continuing coincidences, there were also some more equally fascinating ones. And just one more I will mention that occurred', and that had had a similar startling effect', was when after having started reading a book I had bought a year earlier; that I had picked up, in a reduced price section in a book store. And, remembering that I had bought it, had now decided to start reading it; and by the third night had got through about fifty pages. And also during this evening, had read how the person in the book, had been reminiscing about London in years gone by. And had mentioned' a well known London venue in Leicester Square, called the Cafe De Paris. And had said that he had remembered when Marlene Dietrich, had performed there in Cabaret in the 1960s. 'And, it had caught my attention,' as she had been someone I had been an admirer of. And remember being surprised, as I did not know that she had performed there.

But it was the next day, after having got on a bus, and not the number bus I had been waiting for. But it was going near my destination, and although normally I would have waited for the right bus to take me there, I got on this one instead, 'deciding I would walk the rest of the way. Anyway, after I got off the bus and then cut through a street, then came out onto a main road; I walked past some shops, 'one of which was a curio type junk shop'. And as I walked by, I turned my head and glimpsed at it; and saw, placed in the window, an old record album and on which, was a picture of Marlene Dietrich. 'And written on it, was: Marlene Dietrich at the "Cafe De Paris". And, almost being mesmerized by it, I then went over to the window to have a closer look. And at the same time, was fascinated by it. 'This, which' had a lasting effect on me for the rest of the day. But I didn't leave it at that, and during that evening decided, I would go back there the next day and buy the album, to keep as a souvenir; and which I now have.

And so, it was these, as well as my other theory's, that were also clues, filling in the pieces in a jigsaw; that largely pointed to the fact that I was on the right track. And, these it seemed, and which I now took as little reminders, were to keep me going or encourage me, letting me know that he, Sai Baba, was aware of me, and when the time was right he would let me know when to go back to Puttaparthi. As during some of the times when I had prayed at the Sai Baba Centre, I had asked if I was doing the right thing, and had asked for guidance. "And that if the Entity was not meant to be with me, to take it away from me." And what started to happen after the full realization that it was Sai Baba who was responsible for the coincidences, was that during the times I would now pray, I would ask for a sign that he had heard my prayer. And the acknowledgement I would get, would often be in the form of a coincidence, and could either come days later, a week later, or even the next day. But always unexpectedly! And now when they occurred, they took on a special significance, because I now knew there meaning. And besides this; I sometimes sensed I was getting help from him in other ways too. In fact, by other occurrences and things that came my way, 'I was almost certain of it'.

But it had also been during the summer of 2004 that I had seen a programme on the television, on the BBC, that had been about Sai Baba: Referring to him as India's biggest guru. And the programmes theme was to imply that Sai Baba was in some way bogus, and that much of the stuff attributed to him 'was not true, or that he was not genuine'. And it didn't seem to focus on any of his miracles or miraculous abilities, or giving any account of all the people he had helped using them. Or the vast amount of good he has done! Or the various things he had built and provided throughout India for the poorest of the people, on a huge scale, from hospitals, to schools and colleges, as well as piping a water supply to various regions and villages – all for free. Only some acknowledgements from cynics and sceptics that he did seem to possess some paranormal abilities; but seemed nothing more than a programme whose main purpose was character assassination. And although some of what was said was shocking, and even made me question if it was true, even questioning my own belief in him. I also had to keep in mind my own experiences and experience of him. And based on this, one thing I was 'sure about,' was his 'omnipresent, and miraculous supernatural abilities'. And from what I myself had been witness to – he was genuine!

But he himself has said' that there would be people like this, who would try to destroy his reputation. And he has also said that there would be people who would try to kill him, and which there have been attempts to do this. But he has also stated that they would never succeed. "And it was things like this, these coincidences and other things he had done, and the many more books I had read about him, and written by a wide variety of people from around the world, 'that had left little doubt in my mind', that he could do all the things attributed to him, and that he was something special. And that he was what he said he was: an Avatar, "there was too much evidence of this reality." And it was this that kept me going and made me feel that at some point there was going to be an end to this whole situation. 'And therefore', was putting my trust in him. And therefore carried on working and chipping away at the book; overwhelmingly now feeling I was doing the right thing. And it was later that summer during September of 2004', that I started a computer course, just one morning a week, for three hours. And this was to learn the basic stuff, and type, and, which had been organized by my social worker, at a place run by a mental health service. And although I didn't realize it straight away, I soon realized', that when the time was right and I had finished writing the book; that it would be just what I needed to know, in order to type it up myself. As, up until this point, I had planned to find someone else to type it up for me, and was writing it by hand. But after several weeks, and being amazed at what the computers could do, 'decided', that when the time came, I would buy my own computer,' and type it up myself.

But apart from this overwhelming faith in Sai Baba, now born out of both knowledge and direct experience of his powers and Omni-abilities, there had also been times when I had doubted him, and wondered if I was doing the right thing. And this was something that had never fully gone away. As sometimes the suffering I still had to endure, 'the cruelness of it', and frustration', had often felt too much of a burden to bare. 'And at these times, the writing of a book to reveal to the world,' something groundbreaking in its detail', seemed of no importance; it was meaningless. And what might well fascinate the masses; was to me, an everyday normality; 'of which the fascination, had long ago worn very thin'. And at times I felt at breaking point, with suicidal thoughts never far away. 'And an end, and oblivion, 'seemed like a very nice alternative.' And by this point, apart from Margaret Palmer, I did not really talk about my unusual circumstances to anyone anymore, even my family. And as well as these, and any new people I came into contact with, instead, 'projected the illusion,' of normality.

'But behind this facade, my life in some ways, often seemed somewhere between that of a 'Walter Mitty existence,' and being in the twilight zone! And often seeming like I was leading not a double life, but a treble life: "With three 'faces' and 'three realities' to it? 'And as well as somewhat uncertain; 'was often a private hell!' And although the help I was now getting from the Entity during the times I was writing, was now increasing', 'with it having a bigger input. I couldn't understand why it was often causing me so much aggravation in other ways. And as well as the other things I have already spoke of, "there was now also the sleepless nights." 'And this producing of the insomnia: 'had also become a growing feature of life with the Entity; often being kept awake for hours on end, and often trying to sleep in the day to try and make up for it. And which the Entity would often make sarcastic remarks about, or taunt me during these times. And this whole lack of sleep thing, the nights of it stopping me from sleeping, often felt like torture. And the wider impact of this,' was that it made it hard for me to develop a routine of any kind; and which had been difficult anyway. And there were other more minor additional annoyances it was capable of, and which I was now being made aware of, i.e. like producing the effect of a spider or something like it' crawling on my skin, often while I was in bed, and which would often make me jump, or react in some other way. And additionally it would sometimes cause pains in different parts of my body – another form of aggravation,' that I was now more than familiar with too. And, continued to do things to my teeth. And as well as all this,' there was still the sleep walking antics in the flat, 'it tampering with things', which sporadically continued, and which I could do nothing about either; and, was often fed up with the consequences of. And although this itself wasn't one of the worst things, 'it was the accumulative effect of it all,' that often got to me and caused the despair'. And all of which during the later part of 2004, in early December, had caused me to almost abandon the book, and start to make plans to go to India at the next convenient time or opportunity; feeling, like I could not go on like this anymore! And although all the evidence suggested (or seemed to suggest) that by writing the book first, that I was doing the right thing; and because of the input I was now getting from the Entity, seeming to further add weight to this. And, by all the other evidence that seemed to point to this; by now "I had estimated and had felt about 90% sure that I was,' doing the right thing', "and that this is what I was supposed to do?" And although the Entity had never said that it had a contract with me of any kind, or that this is what it was here for! And, during the times that I had asked it, and tried to find out if this was the reason it was with me, 'it would never give me a straight answer, but instead saying things like, "who knows what I'm here for?" There were also the obvious questions i.e. why had it told and revealed to me everything it had? "All the information in all its detail: About the true nature, of what is behind,' and mistakenly regarded as mental illness." But not only told me, but also put me through it all, and leaving me in no doubt about it. 'Or the amazing abilities that these Entities possess,' and how all this was demonstrated on me. Surely there was a reason for all this. I mean, this is what it had gradually seemed like. And, the way it was all done? And additionally to this I had gradually realized that it could just as easily have cured me of everything while I was asleep, without me ever knowing about it. "And it hadn't had to reveal itself either!" 'It had said to me a while back, that if it had not revealed itself: that I would have never known about its presence! And this I have absolutely no doubt about. "I would never have believed, or thought that I was possessed by an 'evil spirit entity', and that it was the source of all my problems: both mental and physical.' This would have seemed too far-fetched?

But besides this weight of evidence that seemed to suggest that I was doing the right thing; there were some evil things that this Entity had done that had made me question, and made me feel that it was a possibility,' that this Entity could be just stringing me along.' 'As these things I feel were not necessary in any way whatsoever.' And, this was part of this remaining ten percent that I was still unsure about; and was often at the back of my mind; and from time to time had troubled me as to why it had done them! And although there was no doubt about what had been revealed to me; 'it was this part,' of this remaining ten percent, which had me thinking: that maybe it was so evil that it was helping me, and at the same time stringing me along, so it could destroy my life, or ruin my reputation. And from what I had seen, from many other people who had been invaded by one of these Entities: this is a 'feature' and their hallmark; "they do destroy peoples' lives." 'And, in order to comprehend that,' you have to fathom evil, and the very word itself 'evil' and realizing its meaning, and it being attributed, 'to that which is without a conscience; and there's the answer. 'And these Entities' are, without

226

a doubt 'evil.' "So, I also knew, to some degree, that this was a gamble too".

And so it was all this, also, these lingering doubts, as well as everything else, and this ongoing burden; that had brought me close to breaking point. And led me during December of 2004,' to start making plans,' to go back to India at the next possible opportunity. And which I decided would be in the New Year, in February. 'And which at this time, was about seven weeks away'. And this would be as part of a group trip: which was advertised in the monthly Sai Baba magazine' that I had subscribed to, and would be leaving in the first week of February, and staying for three weeks.

But although it was conveniently not far off, it was also a bad time of the year: and for many reasons a bad time to start organizing this trip. And also for many reasons' I felt that I needed a bit more time. So in the end, after Christmas had gone by, I decided in January that I would definitely go on their following trip instead, and which I was told would be the first week of June: and exactly four months later than the February trip. And although feeling desperate to get rid of the Entity, I had also felt that because I had now set a definite date to travel there, 'that I could wait that bit longer. And at the same time, I could use the time to get as much of the book done as possible, 'or maybe even finish it'. And at this point, now January 2005, I had felt that I had got half of it done. "But was also aware, that the most important chapters had not yet been written? "And decided that in terms of getting the book completed, or at least making a big dent in it, I would now have to put my foot to the floor, and develop a strategy, that would allow me to be working on it as much as I could under the circumstances – "flat out." And this is what I did. And as the next couple of months went by, so far, all was going to plan. 'And I waited for the advertisement to appear in the monthly magazine: knowing then that it would be the right time to call and put my name down, and start the process of organizing the trip. But in March, after having received the monthly magazine,' had noticed that 'there was no advertisement for the June trip.'

I phoned the organizers, and to my disappointment' was told that there would not be a trip in June this year: and that the next trip would be in September. And which at this point was seven months away. And although I could have opted to go on my own; 'this did not feel like the right thing to do. And I was also now aware that Sai Baba often looked very favourably on large organized groups, and often granted them a group interview. And this' was also a reason why I wanted to travel this time with a group trip. And plus, everything was organized for you. 'So it was a big disappointment to hear that there was no June trip'.

'After telling them of my disappointment, and that I really wanted and had planned to go in June,' they had given me the number of another organization, that might have a planned trip for this same month. 'And although I took down the number', I decided not to call straight away; but wait for a couple of weeks and think about what I should do. And another reason for this,' was that I didn't have a good feeling about going there, with this other organization. Something about it didn't feel right. And the one that had been advertised in the magazine, this had felt like the right one to go with.

But to my surprise,' it was about three weeks later, and I had still not made a decision on what to do, and, had just received the next monthly magazine. And although I had been told there next trip would be in September, 'at this stage they had not clarified the exact date.' But now when I looked at this month's magazine, I saw their advertisement, and with the date for the September trip; and it was the same day I had travelled there on the last time – the 11th of September. But at this time, and although it was a coincidence, I had still not planned to wait for this September trip, still feeling it was too far away; and, was also still somewhat in a dilemma of what I should do?

But over the next few days, as I thought more about it, I had come to the decision that because things had not been too bad,' and I had been making good progress with the book; that I would 'wait these extra few months!' 'And that I 'would', now go on the September trip instead.' "And at this present time it was six months away, 'and, would be an extra three months to wait'. And, one other thing, that had made up and been part of my decision', was the still lingering feeling in the back of my mind, that if I was right, and there had been a contract of some-kind between the Entity and me, and I was meant to write this book, then these extra few months could ensure that I would achieve this. And, this with the help and expertise of the Entity: whose help and input, was ever increasing, into all phases and aspects, of the process, 'that writing a book entails.

And although I knew it would entail more of the frustrations and ups and downs that were all part of the situation. I also felt that because I had now made a concrete decision and had a date and deadline to aim for, and also a purpose; that I could, or would endure it, for these reasons. "And I could then go to India, with the peace of mind, knowing that if I was right and this is what I was meant to do, and had to go through, or even what I had come into the world for, then I have done it, and then there was no excuse or reason anymore that this Entity needed to be with me. And I could go there, or even stay there, until I had achieved the aim, 'of the coveted interview'. And during which, as well as the aim of asking Sai Baba to remove the Entity, I also wanted to find out if I was right: "And also ask a number of other questions!" And now that I had made this decision, I also felt that something about my decision had felt right. Simply put, it just felt like the right decision. And the more I had thought about this during the weeks that followed; I couldn't help but feel that it was probably 'no coincidence' that I would again end up travelling on the same day as I had the last time. "And, couldn't help but feel that Sai Baba had had a hand in this;" and in fact, had eventually concluded that it was no coincidence at all, 'but just another reminder of the ways Sai Baba works or does things'. 'And after all', I had been asking him for guidance on what I should do about the dilemma of my situation; and I think the answer had come in a way that I was now more than familiar with, "a coincidence" i.e. this was the right time to go – or the best time to go.

So, I now had my date to work towards! And I knew the work ahead was going to be a hard slog! And in the circumstances, was easier said than done? But I now felt that the right decision had been made; or been made for me, and apart from the task of getting this work completed, and which I knew the Entity's help was crucial in finishing these main chapters: knowing they would need to be articulated in the specific detail that was required, 'and "which the

Entity's knowledge would be crucial!" And for this reason also that I had felt I would carry on and endure the continuing scenario until September. Knowing that the importance of what I had to reveal to the world – for this, it was worth enduring. And that it was for a noble purpose; and was therefore, a noble task.

"But another aspect of this whole scenario of writing this book and revealing such a horror story about something that has plagued humanity, and continues to do so, is that it would not be any good if there was no way of being able to identify those who are possessed by these Entities, 'and then similarly get rid of them.' And this was something I had often thought about, thinking that there needs to be something developed which can identify and clearly show them up, and equally, remove them from those that have fallen victim to them. And these were also questions that I wanted to ask Sai Baba. Simply put: would it be possible to develop a scanner type of machine, 'a full body scanner that could show up their presence as clear as an x-ray shows up a bone. And also a second device that after their presence had been identified could cause a reaction to prod them out of the person's body.' And these devices could work hand in hand. As one would be no good without the other. Because the scanner would be used to not only identify their presence but to also make sure they had been removed by the prodding device.

"And it is the development of this type of machinery, which is the only way to ensure a 100% that they have been removed and the possessed individual is completely free from them!"

I myself, believe these types of things are possible to invent. For instance, there are already types of devices used for scanning purposes. Some of which are very sophisticated and can now clearly show up all the organs in the human body. As well as there being other imaging type equipment, like heat scanners, known as thermal imaging which are used for various things, and which show up the heat emanating from something or somebody in the form of a colour: the different temperatures of heat showing up different colours! And which clearly show that energies do have a colour. And just one of the various uses this technology is now put to, is in the tracking of suspects in police helicopters at night.

"And as I already mentioned there is also the photographic equipment that is now able to photograph and show up a person's energy field – the 'Aura'. So it is only a further development of these physics and components, 'or a blending and finer tuning of these types of machines,' that could clearly show up the presence of one of these Entities: As the Entity's too will have a certain compositional makeup, and which will also give off a colour and form. "And, I have absolutely no doubt, that this technology is possible to develop!"

And as for the development of a prodding type device that could jolt these Entity's out of the body? "This idea had come about and had grown out of the theory I had heard: that they might be able to be got rid of by electrical current". 'And whether this holds any real weight I don't yet know', and is one of my intended questions to ask Sai Baba!' And another theory on this same idea of the prodding device had grown also from the realization that if Sai Baba can remove these Entities, then whatever his way or methods of doing this was, it would probably comprise of the use of some types of unseen energy's. And it would just be a matter of finding out what that was, and how to replicate that and harness it into a workable machine' that would produce the same effect. 'And it was this also; that my theory of the possibility of a prodding type device had further grown from'.

But there were other things I wanted to ask Sai Baba as well; and these were questions that the Entity had not yet answered. And one of these questions was – how did these Entities' get in and possess a person? Because, when I had occasionally asked the Entity about this' it had advised me to ask Sai Baba, saying: that if he was genuine, then he would know and be able to tell me. And also saying: that it would be better to ask him, because if it told me; I would not know if it had told me the whole truth. And its reason for saying this it said, was that, 'everything it has told me about the Entities it had also been able to show me; and therefore could observe that it was the truth, and it could also demonstrate that it was the truth, and was therefore, left in no uncertain terms, 'that it was true.' 'And for these reasons said it would be better to find out from Sai Baba.

"And the only thing the Entity had said about this and how these Entities' got in; is that it was hard to get in, it wasn't easy?"

And besides my own conclusions that Sai Baba was without doubt an Avatar and something very special, the Entity had said sometime before this, that it too believed and was aware that Sai Baba was genuine.

But apart from the not knowing in concrete terms how these Entities got into the body of a person; I had by now accumulated a lot of knowledge around this subject regarding this issue. 'And the general consensus of thought, that I had come across, was that people who become victims of possession, or vulnerable to it, 'it had been because of the state of their 'Aura'. This is what I had kept hearing. And as I had mentioned in the last chapter that the Aura had another purpose, in that its makeup was not only an indicator of character or other things, but it was also a protective barrier: comprised and made up of bands of energy'. "And what I had constantly been reading or hearing, was that when a person's 'Aura' becomes depleted in some way; 'it becomes weak,' or weakened,' and this can cause holes or tears to appear in it. And, in some of the information I had read; it had said that when this occurs,' 'it makes the person susceptible, or vulnerable' to invasion and possession'. Meaning: that's how they get in or work their way in! And it was this also, that I had asked the Entity to clarify, "if it was the truth" and to which it had told me to ask Sai Baba?

"And some of the reasons given for this weakening or depleting of the Aura, were: drug use, constant stress, illness, or continued over consumption of alcohol, or even accidents that cause severe shock; and this last one having an immediate impact on the protective make-up of the Aura.

And although I didn't know a 100% if this was the reason conclusively, and that that' was how they got in; but had concluded, that in many ways it made sense. But also knew for sure, that if this was the case and the truth, then only one of these reasons could have applied to me. 'And that would have been the shock I had sustained after being hit by a car when I was nine'. And as I had been made aware of and concluded some time ago; that the Entity' had been with me before that. So this too was doubtful. 'And this was another question I had asked the Entity on a few occasions – exactly how long had it been with me? And what were the circumstances, how did I become a victim "if this was the case." 'And although I had not received a conclusive answer to this either; the only concrete conclusion, that I could draw from what I now knew; was that, it 'seemed like it had been with me since early childhood'. Or the growing possibility and theory that went even beyond this, and that maybe we came into the world together! I couldn't rule that out either. And it was questions like these' that I wanted to ask Sai Baba about also?

"But whatever the reasons or ways that these Entities get into the body of a person; I couldn't help but feel that it's a cruel world, and a cruel state of affairs; that the human being has such a floor in its very makeup, that in turn makes it vulnerable to invasion by such evil." And this I found hard to fathom, and could only conclude that the world can be a very cruel place.

But as I said about the reasons and need for the development of machines that can detect and show up the presence of these Entities' and similarly remove them. There are also other reasons this needs to be done. And this is to prove conclusively, that what I have revealed in all its detail,' is the truth beyond any doubt. As a removal of these Entities from those who have fallen victim to them, would also be a removal of all the symptoms and trouble they cause; these symptoms would no longer exist?

"And to prove also, to those that are involved in the professions to do with mental health; and that they are left in no doubt that this is the truth, and that: This bizarre reality is what they're actually dealing with. Because at present the general line of thinking or that is taught by the Medical Establishment and its affiliated agencies, is that the voices' and other symptoms that are the work of these Entities', couldn't be anything other, 'than a medical malfunction of some kind in the brain.' 'And that the very idea that it could be anything else', is preposterous, or unthinkable, and especially, an evil spirit. But as I have spelt out and made very clear, that those who fall into this category and who are in fact possessed by one of these Entities, display a group of symptoms that sets them apart and clearly distinguishes them, 'from others that fall under the banner of schizophrenia.' And most notable, and the largest and clearest' indication of invasion' by this type of Entity as I have already said, "are the voices." This is the biggest giveaway and instant way of diagnosing the category of the person. **And I write this following section also with the Medical and Psychiatric Establishment in mind.** But as the voices themselves are the largest clearest sign, they are not the only ones; and it will soon become clear once the surface has been scratched that they are just one component in a group of symptoms and problems that these Entities' cause and that the person is suffering from; and can range from hallucinations, to delusions, delusional behaviour, paranoia, strange smells, unusual medical problems, 'that are hard to treat or diagnose or are unresponsive to medicine.' The often laughing out loud, bursting into laughter I mentioned! Disturbing or evil thoughts and images' being floated or projected into the mind. Obsessive and extreme thinking around subjects,' that may bother the person, or that the person is uncomfortable with. Insomnia, anxiety, depression and mood swings, creating them in the same way they can tamper with the body, they can also do it with the brain. And additionally to all this can adjust and tamper with the brains chemistry in ways to cause different states of mind: from a tranquillizing effect, to a severe feeling of being drained, to being severely depressed, to memory loss in the form of suddenly blanking the mind, as well as suddenly make you feel very tired. 'And then,' being able to keep you in any of these states for any amount of time'. These were additional things that were also demonstrated on me time and time again as proof; until I was left in no doubt that these too could be flicked on and off like a switch. And any of these symptoms: "as well as the voices." 'Or all, 'or any mixture of them: "as well as the voices", would be a clear sign and diagnosis', of what I have categorized and have now termed "Possession Schizophrenia." And for the reasons I have already stated. "And it is this additional diagnosis", that now needs to be incorporated into the medical process, of "diagnosing a patient", who is suffering the effects of what I have outlined.

And so this, the above, is a clear indication, a guideline, and a blueprint for a diagnosis that can now be identified as: Possession-Schizophrenia!

"And an additional element that should also be noted; and which had recently been brought to my attention.' During 2006 I had read an article about the development and invention by scientists at Surrey University, of a computer programme', that can diagnose people with schizophrenia. And stating it did this by measuring the level of the Amino Acid N-Acetylaspartate in the Thalamus area of the brain. And which levels of are found to be lower in schizophrenics: And saying that they had tested it on 18 volunteers' half of which had the condition; 'and found that it was 100% accurate.'

But, although the development of this is further progress for modern medicine; and would further assist in a diagnosing process. 'However, it must also be taken into account the 'ability of these Entities' to adjust, alter, and tamper with the chemistry of the brain, and the brain itself. And because of this, the final deciding factor for a specifically accurate diagnosis, of whether it is due to a genuine brain disorder: or is the "Possession" Schizophrenia, 'should be based entirely on the symptoms; and what I have specifically outlined'.

And although I have outlined a blueprint and template for a 100% diagnosis of the above mentioned: 'Possession-Schizophrenia.' There is also another case scenario, in which a person maybe experiencing some, or all of the above symptoms, 'but without the voices!' 'And all these symptoms together', can also be another indicator of invasion by this type of Entity. And although the person may not yet be experiencing voices, these may appear at a later stage, or may not. 'This depends entirely on the Entity'. And, from what I have been told,' and also learned at the hearing voices group, and from other sources of information is that 'many of the people were having other symptoms before the voices came on the scene'. 'And, as was mentioned in the revelations chapter', that although they are all the same type of Entity, with the same basic nature, and ability's, "their range in personalities" is as diverse as that of human beings,' and this is what accounts' for the different extremes of symptoms: And also whether the voices are produced or not! "And that at the lesser more minor end of the scale, some will just live in the body of the person, and cause only minimal problems'. And produce only minimal symptoms?   But at the same time, still affecting, or influencing, or manipulating and deluding: as well as affecting relationships and dealings with people in general, as well as other things: Causing physical health problems etc. And others, at the worst end of the scale, will cause havoc and misery and will delude the person, into doing the worst kinds of things, including murder. 'Or even drive there victim to suicide!

And so, this information of all these same symptoms, without the voices, is a secondary way that can also be used to try and establish if the patient fits into this bracket. 'And it must also be noted, as it is generally known that some of these same symptoms without the voices', "like the hallucinations or paranoia." These on their own, can also be attributed to other things – from drug use to severe sleep deprivation; and which usually subside when these things are no longer a factor. And there are also the genuine cases of psychotic type illnesses and brain disorders, brought on by or due to other factors, 'and which usually respond to medication or gradually get better.' 'But, when these symptoms' are an ongoing long term problem and are part of a wider group of symptoms that I have outlined; it is then that this possibility of possession, "invasion by this type of Entity", should be taken into account.

'And all of these secondary additional factors: 'are also more reasons why the technology needs to be developed, to spot these Entities,' and show up their presence!

And the one last scenario! And as some people would have already heard: Historically there has also been incident's of hearing a voice in the form of a message, and it having a divine significance – something that has been encountered throughout history. But this, and unless it's an ongoing thing' and with negative symptoms, it is not to be confused with any of this; and is of a subject that is entirely different.

And I myself, as well as being put through all of the symptoms and full range of the abilities of these Entities' to the point of being left in no doubt, about what has been revealed to me, have also witnessed close at hand, and listened to many others,' that are suffering and enduring the full range of misery that these Entity's are capable of inflicting on their victim: from the mischievous to the cruel to the evil. And from what I have seen and heard it has left me in no doubt that what has been revealed to me is the truth 'and that this is a Specific Type of Spirit Entity'; and which all have the same character and same abilities, they all create the same types of problems, and they all torment their victim in similar ways: 'And either to a lesser – or larger degree'.

'And during the many times I had been to the hearing voices group', or other times I had come across people with this same problem. "I had heard many accounts of stories by these people, of how they are affected by having this problem. 'And as I said,' the pattern of symptoms and character of the voices are all the same. And although I could give many examples of the cruel or mischievous nature of what I have seen and heard, "I will give just a few examples, of the way in which the victim is often tormented, or generally annoyed – just in the form of the voices alone."

A small but good example of just the lesser mischievous and annoying side of them is when a person at the hearing voices group had been asked what the voices had been saying to him? And had replied: that when he was coming up the stairs to his flat, the voice had said, 'go and tell your neighbour to fuck off!

'And this type of mischievous statement' was typical of the character of these Entities, and it was just one small typical example, of just the lesser more mischievous average day to day types of things that are said to the victim. 'And as well as the nature of things that are said', there were also examples of the fact that the reality of this problem was anything but mental illness? 'But because it's taught that the voices are a symptom of mental illness, 'and which are referred to in medical terms as 'auditory hallucinations,' the clues to this are overlooked or dismissed, and the person is thought to be suffering from a brain disorder: and what is generally referred to as schizophrenia! "And therefore, the elements that are clues to a live personality behind the voices are overlooked," and not taken seriously. And one good example that stood out for all to see was when a woman in the hearing voices group had been explaining how she had been'. "Saying: that she had not been too good, and that the voices had been really tormenting her and saying nasty things to her 'and about her family. And she was asked by the group facilitator, what she did or how she had dealt with it. And she said 'I was very upset and just said to the voice, why are you doing this to me? "And the voice replied – because you deserve it!"

And although this was unusual in the sense that almost overwhelmingly the Entities don't usually answer them back; 'preferring to keep their victim in the dark', and torment them in this way as well. It was nevertheless, a good example of the difference in personalities of these Entities', 'and at the same time, a clue that would go unnoticed: and just dismissed as that person being a bit more ill than usual, or going through a bad phase, or maybe she had been under some stress that week that had made the voices and her symptoms worse; as the general consensus of thinking that is now taught' is that 'stress', makes the symptoms of schizophrenia worse? 'And, although 'this may be a factor' for those that have a genuine psychotic type brain disorder of some-sort.' "It has absolutely no bearing" on those that are the possessed kind! But at the same time something that these Entities will play around with, and use in the same way they do with the medication, leaving all those involved thinking that their right; and that stress does make the voices worse.

"And there was another example: that was even closer to the bone. 'And this had been during the time when I had still been going to the hearing voices group almost every week'. And during one of these times,' a man that I had not seen before had come to the group. 'And although I wasn't exactly sure where he was from, 'I thought by his appearance and accent that he was Turkish. And although his English was not that good; he was managing to explain his symptoms', and how they were troubling him 'in broken English.' 'And the general impression that he gave off, was that he was very troubled and distressed by it all. "But what really stood out for me was what he had said." And he had not only talked about the voices and the things they said; but then described other things that were going on. Saying: (quote) and he causes pain in my leg, and sometime he talk through my mouth as well as say things to me in my head. And what had first struck me was that he was describing it as something more than an illness; 'like there was a live personality behind it all.' "And which there was!" 'But at the same time he still didn't know what it was?' But what had really hit home and horrified me', was that he was sitting right next to me "and that I knew the truth and the reality of his situation". And at the same time: 'was also aware that there was such an evil presence right next to me!'

And although I had often thought and was aware of the fact, that when I was in the hearing voices group, I was also in the presence of more of these Entities', and was observing them and their behaviour from a unique position.' 'And sometimes even wondered if they communicated with each other' and had a good laugh?' 'But none of them had horrified me, and fully brought home their presence in the way that was expressed through this person'. And I also couldn't help but feel sorry for him: Thinking that his situation was at the worse end of the scale. And although I knew exactly what was wrong with him, "the explaining of these additional symptoms to those working in the mental health professions" would have just been dismissed, or lumped in with the rest of his symptoms: the voices etc, and thought to have been of a delusionary nature. 'Or like I said before', the thinking that the worsening delusionary symptoms may have been brought on by added stress, or a particular bad bout of it.

'And these three examples;' are just some of the countless examples that I could mention from what I have seen and heard myself, or come across in articles or other forms of information; comprising written accounts of people describing their symptoms, telling their stories, and how they cope with it. And I remember when I went to the Hearing Voices Conference, and attended one of the talks, about 'coping strategies;' and which the assembled number of people, at this one talk and lecture was around a hundred. And the person, who conducted this, was himself a facilitator of a voice hearers group. 'And, as well as the lecturing and question and answer part of this,' he had also got three men from his group, to stand up in front of those assembled, and read out brief written accounts of their own suffering, and history of it. 'And to anyone sat there listening;' it was grim to hear. 'But everything they said and described,' was what I would now call, "text book invasion and possession", 'by one of these Entities'. But what was even grimmer,' was that the many assembled professionals,' were being taught and fed the illusion' that this was a mental illness, and that those suffering from it, "could be trained like circus animals!"

But having said that: The other aspect is that the hearing voices groups, of which there are a network of up and down the country, are a good thing; in the sense that there is some support for those who have unknowingly fallen victim to these evil Entities. 'And is somewhere to go to be able to talk about it with others that are in the same boat'. And because, 'those' that are enduring the effects of this situation, 'are themselves', in a unique category of their own!' "And often these Entities like to isolate their victims too." And, until there is a concrete way of removing the Entities from them; they offer a much needed crutch', that is beneficial to the sufferer; as I too had benefitted from this. But one thing is certain, and that will not go away; regardless of how many years a person attends a hearing voices group. "And that is, the unpredictability that invasion by this type of Entity brings." And although for some, life may improve in some ways, due to the victim learning to cope with it better, or just accepting the problem more. But, there will still always be problems! And long term; the only thing predictable for the victim, 'is the unpredictable,' and their life will never be normal! The Entity will still subtly manipulate them, influence them, and play their games; and in varying degrees, control them.

"But besides the harrowing reality that I have brought to light, there was an additional realization that had been gradually realized. And that was; that at the time when first given the name, 'The Unwanted Companion', I had thought it perfect for my situation and a perfect name for the book, because that's exactly what it is. But as time went on, and I had met many people unknowingly possessed by these Entities; and had seen firsthand the havoc and disruption and trouble they cause; tormenting, and ruining people's lives. I had realized,' that the name 'the unwanted companion,' was for them also. And that they too, have their unwanted companion.

And so, it was for many reasons that I felt I would carry on and further endure my own suffering and unpredictability in order to tell this story; and the truth and reality behind the voices: 'the real Horror Story! And knowing that everything I have revealed and said about the nature of this problem is the absolute truth. And for all the research, bafflement, misdiagnosis, or various beliefs as to the true nature of the voices: and to the question "it is not known exactly what causes the voices", this is the missing piece in the jigsaw, this is it. Having lived and experienced the full force of this reality. I have lived and suffered and endured the worst and most terrifying aspects of it. I have been shown, put through, and had demonstrated, and pointed out to me, and observed proof beyond any doubt "that this is the truth", a hundred and a million percent. 'And to those who would doubt, or somehow dismiss,' what has been revealed in this book; regardless of how incomprehensible it may be; I say this. I'm afraid it's true, this is the reality. "This is the truth."

Chapter Eleven

Return to Puttaparthi

Now several months later: August 2005.

I had stuck to my timetable and kept my foot to the floor. The weeks and months went by and the remaining chapters of the book were being written, and the task was getting done. And in general seemed to be a less troublesome phase – the Entity' not causing me too much trouble. And the help I now received from the Entity had increased almost twofold: the input' often amounting to forty to fifty percent of the work.

And this further input and help had encompassed many areas and aspects, ranging from advice, to the content of the book, the organizing of it, and the making of notes for the various chapters; and to the actual sometimes feeding information or dialogue to me, during the times while I was at work writing it. And this could be anything from a sentence' to a paragraph, to sometimes half a page or more; or often just a single word or change of a word. And often, when I would get stuck' it would soon let me know where I needed to go next. 'And all these elements were often comprised with its brilliant displays of cleverness, intelligence, and creativity: whether this was helping me write out some specific information, or pointing out a better way to explain something. And it had also started suggesting names for the chapters I had not yet written. And although I had had names in mind for some of them; the ones suggested by the Entity had seemed better, or fitted better. 'And this had first started with the chapter, the Magical Whirlwind. And which at the time I had had another name lined up for, but hadn't been entirely happy with. And then the name for every chapter since then, from the Magical Whirlwind, to Return to Puttaparthi' had been the product of the Entity.

And this help in these various ways did not only occur when I was writing, but also at other times. 'Often during the night while I lay in bed I would receive good advice on areas that I needed to cover next, or be given further bits of information that were relevant, 'or further explained points of importance. 'And which I would now often find myself having to turn the lamp on, to write the information down' so as not to forget it; 'as I was soon realizing, that the information was too important to not make a note of. 'And this could be anything! From something that needed to be explained better, or had been left out of a particular chapter, 'or, needed to be put in.' 'Or from where, or at what point to end a chapter and begin a new one – as well as giving me tips on how to do things more efficiently'. And by now, I had been made fully aware of what the Entity had said some time back: That, 'I would not have been able to write the book as well, without its help!' And this was something I had now realized more and more. 'And for all the resentment' I often felt about my situation; I was more than glad that I had the Entity's input'. And the fact that I was receiving all this help and knew that the help was there,' had made the job easier and more bearable.

And although I had never actually asked the Entity' to help me with the book, or hadn't asked for help during the times when I was writing.' But it was now the case; that whenever I sat down to write, its help was automatic, 'and came of its own accord!' 'And the often game playing', mischievous, or nasty side of the Entity was more or less put on hold; 'almost like it went into a different mode.' And although I had at first started off doing ten pages at a time, and then giving a copy to Margaret Palmer' for safe keeping. I had, by the time I finished the Magical Whirlwind chapter, felt confident enough, to now finish a whole chapter before I gave a copy to her: Now feeling confident' that the Entity would not destroy it in the night or in other ways. "And as well as all of this, it was also getting done at a faster rate. And I was becoming more efficient and more organized." 'And the amount of time being put into it, had also played a large part, as I was now working on it almost every day,' or as much as I could under the circumstances: pretty much whenever I could, and very often at odd hours, late at night or the early hours of the morning etc. And, my determination and desire to complete the task, and get all the necessary chapters' and specific information in them done in time, had pretty much taken over my life.' And where the year had begun in January, I had put my foot to the floor, 'and through many trying times' had kept it firmly in place; "and it was now the middle of August," and I was just three weeks away from my long term objective. 'And that was to complete the task of writing the book,' and to return to Puttaparthi.' And hopefully then, 'the mission I was by now overwhelmingly feeling was part of my reason for coming into the world, was then completed and put down in writing, and of which, hopefully there would only be one chapter left to complete. And that was whatever happened next!

'And at this time, during these three weeks before this trip back to India, I had now almost completed this task: 'which I had set out almost three years earlier.' And had now,' got every chapter, that was a necessary part and parcel of this story completed. And was now at the beginning of what I had assessed to be the final chapter: "Return to Puttaparthi."

"And, although these three years had taken their toll; and had put me through the whole range' of emotions and frustrations. 'They had also been' a great growth period!" And I had learned to look at my situation from other perspectives. As, all that had happened' had now changed the way I viewed the world. And one of these had been a result of the 'culmination of information', on a subject, that I had by now, read so much about'. And had largely concluded, was probably tied in with my whole situation? 'And this was the subject, of "Karma and Reincarnation." And, as I had first mentioned at the beginning of the last chapter – was no longer something I could dismiss! "And had to at least look into this aspect, of a 'teaching', that is a core belief, and central component, of religions,' 'comprising' almost two thirds of the world's population?" And from other sources of information I had come across, I had also found out it had also been a part of the early teachings of Christianity! 'But, had been deleted, from the Gospels and Scriptures that had formed what we now know as the New Testament.'

And, although I had come across varying accounts as to why this had happened,' and these teachings were 'removed', and no longer a part of the teachings of Christianity. With some accounts saying', there removal had been instigated in the 4th century, in 325 at the 'Council of Nicea,' by the first Christian Roman Emperor 'Constantine'. Citing, for political reasons: as well as to do with unity in the church. "Because of the different competing sects' and factions of Christianity, that had existed at that time!" 'Often differing in what was being taught', and was often the source of conflict. And this Council, which he had overseen', had resulted in establishing what is known now as the Nicene Creed: The first uniform Christian doctrine; "and, 'apparently along with other things, was when they were deleted!"

But also,' learning that this, 'karma and reincarnation,' had also been a part of the early Christian teachings, had only further deepened my curiosity. And as well as this, there are other historical accounts of further later' attacks on these teachings, as they didn't just go away overnight; and had still comprised a part of the teachings of many Bishops and Priests long after this.

So in my search for answers, to do with my own situation, these are some of the things I had come across myself. And,' whatever the historical facts are concerning the disappearance of what was once a core part, of the teachings of the early Christian church; or all the specifics as to the reasons why. 'The real discovery and proof that they had existed in the first place, is in the remnants, which were left in the New Testament.' And to anyone with some knowledge and understanding' of the wider teachings of karma and reincarnation, they are very visible and clear to see; and understand there meaning. "And although these core teachings (and their significance) are no longer present in full, I had also found out that they can still be found through other sources. 'One of which, is the "Aquarian Gospel of Jesus the Christ?"

And just two of these existing remnants that are still present and that I will mention, and which point to and give a clue to the wider content of knowledge from which it was part of, and had originated from, are firstly: a reference to reincarnation and past lives', which is referred to during the healing of a blind man, in the Gospel of John, John 9: When Jesus and his disciples come across a man who has been blind since birth. And the disciples ask Jesus: Rabbi, who sinned, this man or his parents that he was born blind? Obviously: referring to past lives', and the role of karma and reincarnation.

And although the biggest component of these teachings and knowledge that had been lost, was this reincarnation aspect', and what Jesus Christ had taught about this. The second remnant and reference, and other part of this whole component of knowledge, that works alongside it, was its other aspect! And which is commonly known and referred to as karma. And it is this, which simply put, means, "you reap what you sow," (good or bad) – action follows reaction

And it is this, 'that is another hint and remnant of these teachings' that still remains in the New Testament. And of which most Christians are familiar with. And this being' this simple statement said by Jesus: 'A man reaps what he sows.' 'And it is this, simple statement,' that is karma through and through'. And which alludes and applies not only to having good or bad actions that are done in the present lifetime, coming back to the person in some or a similar form during the lifetime. "But also incorporating those," which are of a nature that could not be fully or completely balanced out in the previous life: or this present life. These would continue over into the next' or future lives, until fully balanced out.

So, in fact, stating that, although we may think that we get away with bad, dishonest, or criminal things, we actually don't. (We get away with nothing)! And this too is what was being referred to when he said, a man reaps what he sows: As Ye Sow So Shall Ye Reap. It was being conveyed and referred to in the full context of the wider content of his teachings, and this wider role of karma and reincarnation at work together.

And although the main teachings of the reincarnation aspect, and of reincarnation in general, are that we as a spirit-soul keep reincarnating into flesh in order to evolve and perfect ourselves through experience, so that we can eventually escape from the cycles of birth, death, and rebirth, and no longer need to come back anymore. 'And when, this evolutionary stage', has been obtained and accomplished through experience; we can then go back to 'from where we first came from', and from all accounts: where we can eternally live in bliss.

And although I've put these in very simplistic terms' and that there is a lot more detail and knowledge on the various stages and aspects of this process, this is, basically what it is and what it says. In short, it says our time here on this earth is where we learn and evolve, and not our real home.

"And the reason we have no direct recollection of previous lives is deliberate. Because of the baggage or burden, and problems this would create or cause if we did know – the burden would be too heavy. But although there is no direct memory of past lives; the effect of previous lives lived, seems to be that an imprint, 'an impression is left.

"And the element that works hand in hand with this evolutionary process and journey of our spirit,' is this karmic karma aspect of: As ye sow so shall ye reap!" 'And although this karma aspect may seem or appear like revenge, or pay back; from what I had gradually realized is that its true role,' is one of a balancing out nature: 'so that we can move forward, make progress, and learn through the various stages of our evolution. So in effect, saying, that not only could we be here with a blueprint incorporating some karma from a previous life; but can also additionally accrue negative or positive karma while we are here! And, that although a blueprint for our life will be in place, and that at certain points or stages as the lifetime unfolds', these aspects will take hold and present themselves. Additionally to all of this, is the reality that as human beings we are also given free will. 'And it is this free will,' that can also get us into more trouble along the way; often causing us to accrue the more negative karma.' And which most of will come back to us in some form, or a similar form, with a similar theme, during our current lifetime.

"But as a further guide to us, in order to help us not to bring this about, 'we as human beings are also given a "conscience" (And therefore a choice).' 'And of which it is said to be the God that runs through us all: "The spark of divinity within each of us." 'And in the sense that when the promptings of the conscience is listened to, will tell us right from wrong, good from bad; and even help us to make the right decisions. 'And a further example of this', is the meaning of the word it originates from, 'which is the Latin', "Conscientia," meaning: 'knowledge within oneself.' And which,' also further illustrates, and implies, that although the conscience is part of our makeup; it is also a separate body,' with its own intelligence, and moral code. 'And it is when we go against this: act against the dictates of our conscience: consciously know when we are doing wrong and do it anyway, (an example of our inherent free will), that bad things will follow.

But I now, had not only recognized and come to the realization, that all of which I had by now read and heard about karma and reincarnation, seemed to make sense. And seemed the only thing,' that gave any real logical answers,' to all of the many unanswered questions, 'to do with what often appears to be the unfairness and unevenness of life. But had also gradually recognised that it was the only thing that gave clearer answers to the deeper questions to do with life, and our reason for being here. 'And seemed to be the only thing that made sense of it all? And even many years before I had ever known, or heard, or read anything' about karma and reincarnation; I had at least been aware of something at work in the world that seemed to hint at the reality of this karmic, 'reap what you sow aspect'. And had observed this at work in all manner of ways', seeing it being perpetrated back to people for things they had done in their present lifetime – the fruits of their activity's coming back to them. And whether these were people I have known, to historical figures, or famous people, or to people you hear or read about in the news or newspapers. I had observed what would generally be termed: as getting their comeuppance. And although their comeuppance is what this may seem; fundamentally it is karma at work; natural justice; the fundamental principal of you reap what you sow, very simply: "The balancing out!

And from what I have seen myself, it does appear that the negative – bad karma, that is accrued in a person's present lifetime, and which can be balanced out, will be! And whatever cannot be fully balanced, due to its nature, will be carried over into the next or future lives. And these it seems, often comprise of the more extreme or horrific or complex things that would take another, or more lifetimes to complete this process. But what this also confirms; 'is that, just like the invisible mysterious forces, that suspend, animate, and regulate our planet and the universe. It also seems that there are equally invisible mysterious forces and mechanisms at work; comprising of an all seeing, knowing, and recording structure. And that propel the workings of this karmic balancing out process. And it appears that it's a natural a part of, and component, just as are all the other elements that regulate our existence.

And it wasn't until the summer of 2005, that after all the information I had come across and numerous books I had by now read. And which were from various sources, not just religious; as well, as what Sai Baba himself has said.' And much reflection, on this subject of karma and reincarnation! And the gradually recognizing and observing of this bigger picture at work in the world. That on one day in particular around this time, it had finally and completely dawned on me, in quite a profound way. 'That this core teaching of karma and reincarnation,' was the only thing, that completely, made any real sense, and gave clear logic to everything! "Seeming, almost like a revelation!" 'Almost like a fog had cleared, and the workings and effect, was there to see!' And, as there is: and can only be one truth. Based on the collective visible evidence, 'it now seemed to me that this, was probably the one truth,' and fundamental reality.' And after all, as well as all the other things and reasons, it didn't seem so hard to fathom, that if the spirit-soul could leave the body at death and go somewhere else; then why couldn't it come back in another?

And so, it was this: this karma and reincarnation aspect' that I had now also pondered over. 'And as well as it now feeling like a frightening reality,' I was also wondering, what aspect of it was responsible for my own situation. And, by now, 'I had one theory of a possible three,' or most certainly,' a 'possible two.' And which also seemed to fit in: "with my overall theory," for my predicament. And this first theory; was what I had heard and read about the different evolutionary stages that we as a -spirit soul' go through. First starting out; 'and through our evolution, over many lifetimes', gradually becoming a young, then middle aged soul, and eventually and old soul; gaining the depth strength and wisdom through experience, and reaching a state of evolution, and spiritual maturity, and where we no longer, or: 'have to incarnate', for ordinary reasons anymore. 'And reasons for coming back at this stage,' can be and often are to take on tough tasks; usually of an altruistic nature: 'and reasons in order to do something for humanity.' And these are often the hardest and toughest tasks. 'And which can encompass many things.' And these people are known as old or evolved souls: "Essentially, now on the spiritual path," and almost at the end of their journey!

And as well as having often felt, like what I now knew was termed an old soul, and instinctively recognising this aspect, and it resonating with my own sense of self. And, it answering many other long held thoughts and questions as to how I saw myself in the world, and often felt amongst people. 'As well as my sense and awareness of things since a young age' that other people in general,' seemed to be oblivious to. 'And it was this aspect,' of karma and Reincarnation, that also fitted in, and resonated with my theory' as to the reasons' of what, and why, I was living with', 'and what' I was having to go through – why I was in this position!" 'Going hand in hand,' and seeming to now, 'be another piece in the jigsaw,' and fit in with my overall 90% theory.' 'That this was probably part of my life's purpose: "Or, even the main reason for it?" And 'was also coming round to the theory' that the reasons the Entity may not have given me a straight answer and told me the truth about this, (if I was right) was because if it has done so; "then that would change the whole nature of the relationship.' 'In a sense', that in order for me to be able to tell the story: I first had to be put through it, and shown every aspect of these Entity's abilities, and capabilities, 'as well as character.' "So that there was no doubt in my mind of the truths it had enlightened me to." And that if the Entity had told me this,' and confirmed my theory to be true, and that I was right about its role: 'this would change everything.' And it would then have to stop doing what it was doing.

In short, it also seemed like some, or many of these things done during the last couple of years,' may have been done so that I would concentrate on the writing of the book; 'fuelling me to get on with it.' Meaning, the sooner it was done; and if my overall theory was correct: then the end 'to this situation,' would be brought to an end a lot quicker.

And as well as several indicators' along these lines of thought; and just one being: "The way in which it had stopped me from trying to meet a girlfriend! It seemed very likely, it was done in order so that I would not get distracted by other things, and instead spend my time concentrating on the book, and getting it done -- getting it written.

"And another piece in this growing jigsaw', was the amount of involvement the Entity now had in the writing of this book; and making sure I got it right?" 'And the fact that it no longer caused problems during the times while I was working on it', was another indication.'

But having said that', I still wasn't a 100% certain that I was right either.' And there was still the doubt and the not knowing, of what I had assessed to be in the remaining 'ten percent bracket,' and which, had comprised a number of some still unanswered questions? And which', was also a part of my 'second theory,' and, was nothing to do with anything about, karma and reincarnation. And I had at times, just thought my situation was simply just a case of bad luck! Feeling that I had probably fallen victim to this Entity,' at some-point during my childhood. And in what circumstances I did not know? "And that maybe I was just being strung along?" 'And, this line of thinking' was 'also' a part of this ten percent bracket; 'and which,' was still part of the equation.' And although, only part, 'of this lesser minimal percentage,' could nevertheless, "be the harsh reality". And as of yet; if I was certain of anything; it was that I was certain of nothing!

And then there was the other reality: 'my third theory;' and other aspect,' of this 'Karma and Reincarnation, that had played on my mind? And had made me wonder and question if my being possessed and being put through all that I had by this Entity: "Was the result of negative karma?" 'And had come about,' because of deeds done in the previous life, or lives. "Then if so: "It must have been something terrible? Something very bad!

"And although a few years ago, I would have thought of this as being ridiculous; 'it was now a question, I had to ask myself again and again; even if I didn't like it.' And if it was this, that was the truth', 'then this would also be true for every other person, who had been invaded by one of these evil Entities.' And therefore nothing could be done for them either! It would simply mean that it had been allowed to happen, it was their karma.

"And although this one seemed hard to believe!" 'Whatever the truth' that lay behind my circumstances; and how uncomfortable, bizarre, or harrowing the reality could be!' These questions, had also become part of the list of questions that I was now compiling: in order to ask Sai Baba; as he was what I now perceived to be the key, 'unlocking the answers to all my questions.' And an additional question I had now put on this list was: "Who was I in my last life?"

And so, this: my third theory and what I had once thought of as the ridiculous; had now become a 'possibility,' and a very real part of the equation. And with my list of questions and self imposed task of completing the writing of the book, before the deadline of the '11th of September', now done. I was now once again, preparing for the journey back to Puttaparthi: "And now with only two weeks to go."

And, although still harbouring a certain fear that the Entity could turn very nasty, and might still do something to stop me from making the journey. 'I decided, to just leave it to chance' or fate; feeling that this time', 'I had no choice'. And after all, as I knew only too well, that if the Entity really wanted to stop me from going: 'it could do so in many ways.' "Whether this was to incapacitate me in some way,' or make me seriously ill; or simply induce me into a deep sleep and switch off the alarm clock: which it had done before on a number of occasions'. And that, this time, simply 'protecting my passport,' as I had done the last time,' was no guarantee of anything. So this time I didn't even bother to take the precaution of doing that either!

And although I had known, or at least been aware, during the preparations for my last trip three years earlier; that giving my passport to someone to look after was also no guarantee either: and that the Entity could stop me from going by some of these same means. But my thinking during that time was that protecting my passport 'was at least some guarantee that I could get there, 'whatever the state I was in,' 'and that I would do my utmost' to drag myself on to the aeroplane: "even if I had to be carried on a stretcher;" 'so determined was my state of mind to make the journey.' But now, and although just as determined, I also knew a lot more about the Entity; 'and was more realistic.' And knew, that if the Entity really did want to stop me from going, it could do so; and there was absolutely nothing' I could do about it. And although there was an element of fear that this could happen; overwhelmingly, because of how the situation had further evolved; I sensed that it was more likely that it wouldn't happen; 'and in the meantime, was getting all the necessary things done in time for the trip. "And a week earlier had been to the home of the organizers of it, to meet them and the other people that were also going. And of which, including the couple who were the organizers and would be taking us, there were twelve all together: eight women and four men, and of various age groups. And which this meeting' had been arranged for the purpose, of providing us with additional bits of information: or any questions we would want to ask. And, to generally meet' and get acquainted with the other members of the group.

And now, having got all of the main things done, all I had to do, was wait for the day to arrive. 'And, although I had got the main writing of the book', to the stage where I wanted it. In the two weeks that were left, I had continued to write and do further work on it, and continuing this until the day of the trip. 'And when finally the day arrived,' and after setting off and then arriving at Heathrow Airport and meeting up with the rest of the group, and then later boarding the aeroplane; 'once again, on the 11th of September,' 'I was on my way to Puttaparthi.'

"And, after having gone through all the necessary stuff before boarding the flight, and then finally getting to sit in the seat on the aeroplane, and then it finally taking off; it now just felt good that I was 'finally on the plane', 'and finally on my way.' And if anything could sum up how I felt at this time, it was simply just a sense of relief, and that's all.' A sense of relief, that I had made it this far and everything had gone to plan. And a sense of relief, that I was sitting on the aeroplane and it was flying to India.

And then as I relaxed and the journey unfolded, and the reality of what now lay ahead: with the knowing that I would actually be in Puttaparthi in two days time, and would once again be seeing Sai Baba; I now pondered and reflected on the reality that I now faced; 'and, many thoughts,' had gone through my mind.' "And one of which was something that I had increasingly often thought about!" And this was,' if I achieved my objective and the Entity was removed; I wondered how I would be different,' and what I would be like: minus the Entity's influence; and even what I would look like? "And I recalled and thought about what the Entity had said to me three years earlier; it saying to me, that if it ever left, or was removed – 'that I would be introduced to myself.' And as the gradual discovery of realizing just what influence it had had, and still had on me, and both of these without me knowing and then now knowing. And now aware of its full impact; gradually bringing me to the realization that I was no longer sure of who I was. And then fully recognizing that this was a reality; and had asked myself, who am I, and who was I, and then who was the real me, and what was the real me like minus its influence? "And now," I wanted to know' and feel this real me! And this was something I now felt more than ever. And a further reality' of this Entity being removed' was the realization of a new beginning, a fresh start, and my mind back, free from its influence, and its interference. 'And it was an exciting prospect,' and I hoped and prayed that I would now finally get what I wanted.

"But as well as this, I had also now pondered on the reality' of life without the Entity? And more so to do with the more positive aspects of its presence, and which I have already mentioned and talked about. 'And knowing full well,' 'that these aspects were something I would miss' and would be hard to let go of. And now, 'pondering over this very real prospect,' 'I even felt somewhat fearful,' that I would then be alone; "and without these positive aspects" of life with the Entity. 'And would no longer have its expertise, its insight, its perception, its wisdom or its creativity, or magical methods of communication to call on or experience. Or the knowing that it could cure me of anything: as well as its protective influence. And as I said earlier,' it was some of these and this more positive side of it that had often made it good company. And it was all of this that I would miss.

"And although I knew that this would be hard to let go of! I also knew it was an illusion, to want to keep living my life with this situation. And knew full well' the downside and the duality of it all. And knew deep down; 'that the benefits of living with the Entity, were often outweighed by all the negative things!' 'And it was an illusion to think otherwise.' And certainly no time to be sentimental either! And furthermore, I did not yet even know the true status of my situation, and reason for it.

'But fundamentally;' regardless of what the true status was, 'I was now completely sure of one thing.' And that was that I wanted the Entity to now leave, I wanted it out of my life, I wanted to be free of it. And this was the whole purpose of this journey; 'and which I had long waited for. And as the plane journeyed on and the time went by, and my thoughts eventually turning to other things, or distractions; and the plane later eventually reaching Mumbai and landing; and then us waiting and changing for the final flight to Bangalore. Then' reaching it a few hours later; 'and in which we had hotel accommodation reserved.'

"And, after finally arriving there, in the early hours of the morning," we were soon in our rooms and eager to get some sleep. 'And it would be here,' where we would stay one more night: before travelling on to Puttaparthi. And, it was a nice welcome rest; 'from a journey, that had taken almost 24hours!' And after sleeping until mid-day and then later meeting up in the evening for a group meeting, and then later the whole group having a meal together in the hotel restaurant, it was then bedding down for the night and getting ready for our early morning call, 'and final stage of the journey, by Taxis,' which would be arriving at 7 a.m. 'And it felt good to finally be here in India, and about to make this final stage of the journey'. And as morning arrived and the Taxis arrived, we were soon up and on our way, for the three and a half hour drive to Puttaparthi.

"And as we had set off, on this final stage of the journey: "I had thought more about my unusual situation; and reason for being here; as opposed to the other members of the group. And, I had thought about my fellow travelling companions? "Thinking, that little did they know the secret that I harboured; 'and that the group of twelve;' 'had an additional member', and that there was in fact, thirteen of us; "the Entity," being the thirteenth member' of this entourage.

And then the journey itself seemed to be fast moving; seeming to go by quicker, than the last time I had travelled this route. And before we realized it, 'we were there.'

"This time I would be staying in the Ashram itself; which also has a large complex of buildings, for the purpose of accommodation for devotees. And which, a section of is set aside' for foreign devotees. And it was in one of these buildings that we were later given our rooms in, sharing two to a room. 'And from what I had heard prior to seeing the room, the general impression I had got was that the rooms were small, and very basic. And so I expected the worse. But I didn't care about that; thinking, that as long as there was a bed and somewhere to wash, it would do. And, 'whatever I had to do without' was a minor inconvenience compared to my reason for being here. But when we were given our room and after opening the door and entering the room, to my surprise, it was much bigger and much nicer than I expected, and had everything I needed; and I was very pleased, saying to my fellow group member and roommate: that it was a pleasant surprise and was much better than I expected. 'And it also had a great view of the surrounding mountains.'

'And as well as being pleased with the accommodation,' there was also a sense of relief and a feeling of elation 'that I was finally here.' And that I could now embark on my goal' of getting the interview with Sai Baba. And, by what I had now been told,' about how things had changed since my last visit, and that Sai Baba, was now giving very few interviews; I also knew, the prospects of achieving this were now a lot slimmer'. "But I also kept in mind my unique situation," 'and the fact that he would know' without asking, 'why I was here.' And also reminded myself of, and held on to what I had heard and read about Sai Baba, giving interviews, to those who were most in need of one. And one thing I was sure of, and that was, I fitted into this category a 100%. And, it was not long after arriving here and getting settled in, just three hours later, that the group met up for what would be the afternoon Darshan, and which, Sai Baba was scheduled to appear at three o'clock!

"The rules of being part of this group trip were themselves very relaxed." And the only major requirement that we had to adhere to, was meeting up with the rest of the group every day, for the morning and afternoon Darshan. And which suited me just fine. As this is what I had come here for, and if I was going to get an interview or group interview, 'I would need to be at these.'

'And then after we met up and went through the familiar procedure of sitting and waiting in the lines', and the time, soon passing by; once again, I was back inside the Temple, sitting and waiting for the person I had last seen three years earlier.' And although nothing had changed in the queuing procedure, 'the procedure now being used by Sai Baba', was very different, from what I had seen during my last visit! And which, I had been informed about, and was somewhat prepared for. 'And now' when he did eventually appear, 'instead of him walking about taking letters and engaging with the crowd, like he used to, this time he was slowly driven down the aisles of the Temple in a small car, with his window rolled down: "And which, had been specifically adapted so that the passenger seat which Sai Baba sat in could be wheeled in and out of. And, it would stop at certain points' and letters would be taken from people in this way. "And what I had now also been told: Was that he rarely ever picked people out of the crowd for interviews anymore either." 'And on sitting here, and seeing this new situation at work, it seemed to have made my chances, of getting to even speak to him,' very remote indeed. 'Because as well as him being in the car; when it stopped, to take letters, it was his guards' who he would beckon to take them, 'then pass them onto him through the window.

But it wasn't just this that was a setback for me; 'but also the reason why he was being driven round in this car'. And this was because of an accident he had had two years earlier, 'which had culminated, in him breaking and damaging his hip. And this was something I had heard about during that time; and like many others had later asked the question, and wondered why, with all his amazing powers, why he had not cured himself. And from what I had now gathered' it seemed like his situation had gone from bad to worse! Because when I had last heard about this accident and any developments, 'sometime later,' he had then been coming into the Temple in a golf type buggy, and would still stop and engage with the crowd etc, 'and sometimes still walk a bit'. But now he was being driven in a car and didn't even do this. And on seeing this" on this first day, was uninspiring to say the least. And later after leaving the Temple, it had actually made me rethink; and ponder over the possibility' that there could even be something bogus about him. And, I had to have a hard think about this,' and really weigh up the situation. And, was torn between this, and all the things I had experienced myself – as well as heard and read about him. And regardless of his present state of health' and the reasons for it, I had to keep this in perspective. 'And also try to keep in perspective, the reasons I had been hearing as to why he had not cured himself; "although he himself had said he was in no pain!"

"Hearing, that the reasons for this could be of a teaching nature, for the purpose of his devotees, and maybe to instil faith? And this' I was told was one of the reasons for it. This is what I had heard!

But, to some, who didn't really know much about Sai Baba, their immediate thoughts" would probably be that it was a convenient excuse, or that he was a fake, or conman of some sort. And I too, would have probably had similar thoughts. But, was in a better position,' to make a more accurate assessment, 'and judge the situation on my own experiences; and what I knew about Sai Baba. And another reason for this, was that I had heard and read of other stories in the past, of other times, when he had become severely ill – having strokes and heart attacks and other serious life threatening things: "Himself saying he had endured these because he had taken on the karma of some of his devotees: 'Saying that if they would have been subjected to these ailments, they would not have survived them. And so in order to save them he had taken on their illnesses, and in time, to the amazement of doctors and other medical experts, when it had looked like death was imminent 'he had cured himself of them.'

"But the problem with the present situation', was that it had gone on for much longer! And he had now been in this present state for over two years, and because of this, had left people scratching their heads, as to the real reason, 'behind the current predicament. 'But as one elderly devotee who I had later got talking to, and who had had vast experience of Sai Baba, had said: "you can never be sure, or know the real reasons, as to why he could be doing this." And also gave a couple of his own reasons', as to why he thought he had not fully cured himself. And so, there were any number of reasons, as to why Sai Baba had allowed this to go on.

And although being met with this new situation on the first day had put a depressing dampener on everything: even making me have doubts about him. And which,' after, took me several days to get my senses back to normal, and, start to bring back my faith In him.

And, 'something, which had helped to restore this faith; "was the occurrence of three coincidences. "Which had occurred over the space of the following three days - after this first Darshan?" And although these were the more minor type of coincidences, and not the type that would send a shiver down your spine, 'were none the less,' unusual enough; and seeming to be accompanied by a feeling and a knowingness, and the type, I had now come to recognize' as one of his calling cards. And probably the significance of three of them, occurring over the space of three days, 'I think, was probably engineered, as a way for him to restore my faith.'

But even so, and even with these three coincidences" I was also beginning to wonder if it was possible that the Entity and where ever it came from or was connected to, 'could in some way be capable of working these coincidences too?' And had pondered on this before; thinking: "maybe some of the coincidences were of its doing!" And after all, with all that I had witnessed from the Entity, this seemed a possibility that could in fact be possible. 'As, I still didn't know where it came from. Or if it was from some kind of evil realm in the universe,' with its own power structure' consisting of similar abilities to those of Sai Baba.

And, this was something I had thought of before. "It hadn't been the first time it had crossed my mind."

"But, it was a fourth coincidence that occurred, that had really helped to bring me out of this frame of mind, and end the nagging doubts and raise my mood out of the depression I had gradually sunk into during these first few days. And this was because this one had a special significance, of which Sai Baba would know I was aware of. 'And, it had manifested itself in the form of something someone had written on a five rupee note; 'which had been handed to me in change the day before together with some other notes.' And had noticed then, that there was something written on one of them but did not bother to read what it had said. 'But it was the next day while in my room, and after paying for a delivery of bottles of water,' and then noticing this writing on this five rupee note again. And it was then, that I took an interest in what it said, and read it. And it was a simple statement written in English, saying: Amish weds Aaxti on the 11th September 03. And, what was significant about this', was

"The 11th of September." And it was this', 'the 11th of September,' which had caught my attention; and resonated with me. 'Knowing, that this date had played a role in my two journeys here; and, here it was, popping up again, "and now in the Ashram." And as I stared at it, and partly in awe; and then thought about it; I gradually realized,' 'that this was no accident,' or coincidence as such! 'And, had an overwhelmingly strong feeling; 'that this', had been orchestrated by Sai Baba'. In fact, the more I thought about it, and looked at it, I had no doubt that it was. It was typically Sai Baba through and through. "And, it was no accident; this had fallen into my hands!"

But having said that, it wasn't things like this I needed anymore, 'it was an interview – I needed to speak to him? But the good thing about this,' was that I started to look at the whole situation in a more positive light; and my faith in Sai Baba and his miraculous abilities and who he was, 'was gradually restored.' And, after having already so far attended every Darshan, and forensically observed this new scenario, I was now starting to put together a plan for the remaining two and a half weeks that I would be here. And, as the days were going by, I had, by now' already noticed what I had been told! And that was that Sai Baba now rarely picked people out for interviews anymore? As by the end of this first week here, the only people I had seen being granted an interview: 'were a large group of devotees from Singapore. 'And this I later found out was because their leader was the head of the Singapore Sai Organization.' And you could always spot a group by a certain logo they would all ware, either a silk scarf or emblem of some sort around their neck. Our group had one in the form of a ribbon. So, in terms of being granted an interview; up to now it wasn't looking to good.

'And although I was now generally feeling more positive; my moods and thoughts had still continued to fluctuate. 'And by the end of this first week, after having observed this whole new situation, had now even been contemplating the unthinkable reality; that I could even be going back to Britain in this same way (with the entity)! 'And this thought,' and reality', really horrified me: "Feeling that I was now having to think the unthinkable!" And feeling that I could not endure the burden of this situation much longer; and had thought,' that if this is my karma, and there's nothing that can be done for me; rather than continue to have to live in this way, 'I would rather be dead.' And, the idea of suicide,' was a reality that I was now facing, and, was now giving serious thought to.' And this is what it had come down to. And during these fluctuating moods and feelings: in the days of this first week, this is what I had more or less been planning – that if I had to go back and continue to live in this way, then I would kill myself; fuck all this I thought.

But although feeling like I was having to face up to this harsh reality! "I was still aware it wasn't over yet; 'and there were still two weeks left'. And after thinking about the Singapore group being granted an interview and the five rupee note, it at least gave me a bit of hope; and helped further pull me out of this frame of mind. "In the sense; 'that although it was now rare to be picked out of the crowd,' he still did give some interviews, 'and that there could be a slim chance, of this happening to our group. And, was also keeping in mind' the unique situation I was in.

And thinking along these lines also helped to make me feel a bit more optimistic. "And another glimmer of hope arose towards the end of this first week when his car stopped by us and our group leader's letters were taken. "And now because of this I decided to write a letter myself: "further explaining my desperate and unusual circumstances," and prayed in the Temple asking for him to stop and take it; and saying to do so within the next two days. 'And on the second day his car stopped by us and it was taken from me? And this also gave me a bit of a boost. And in the meantime, over the days following this, I had started making the most of my spare time, by continuing to work on the book – where I'd left off; keeping it up to date, as events were unfolding, 'while it was fresh in my mind.

And also, as the days were going by, and the middle of the second week now approaching, because we had been getting up very early,' and often getting places in the first or second rows, I knew to, that Sai Baba had had a good look at us, and was at the very least' more than aware of us.

But I was by now, also aware that my situation was getting a bit desperate! And realized,' that if I was to even get to speak to him, I was going to have to do something a bit drastic: "And would have to flout the strict protocol!" Feeling by this stage, I had absolutely nothing to lose. And therefore had a planned strategy in place, that I planned to use, and would start at the beginning of the third and last week: if by which time I had not succeeded in gaining an interview by the usual procedure. And by now,' had become determined to not leave here without trying to at least speak to him.

But first, I would wait until the end of the second week to see what would happen; and then that would be the deadline. And then I would embark on my plan. 'And which was to simply wait until the car stopped near me, and then quickly get up and approach the car and then ask him myself. And I had even prayed and told him what I intended to do. And as the week went on, and the deadline approached, and the end of this second week arrived, and nothing happened; my plan was now put in place. And I was now on red alert, 'to strike, when the next opportunity arose.

And the start of this was the following day, which was on the Monday; and passed by, without this opportunity arising.

And so, I was now at the point in the stay where we now only had six days left before we would leave. And up to now I thought I had done everything right; and hadn't missed a single Darshan! Getting up every day at 3.45.a.m. to be there for 5.00, and then wait for him to show, which he now did anytime from 7.30 to 8.30, sometimes even later.' 'And which even this had become unpredictable!' Unlike the last time I visited he was always punctual; arriving at the same time every day.

But nevertheless, in order to get a place at or near the front, it was necessary to have to get up this early and wait this long. As by 7.30 – 8.00 o'clock, the crowd by this time had gradually swelled anywhere from twenty thousand and upwards. And there was no way you would get a place near the front, or even close to one of the aisles. And then for the afternoon Darshan, we would queue up and do the same thing all over again. But one good thing about the afternoon Darshan was that it was usually a much shorter wait. And I had managed all this, despite the often disruption to my sleep by the Entity: 'And often feeling shattered or very tired because of it.'

And so, with my plan now in place, and this first day passing by without the opportunity I was waiting for- it was try again tomorrow.

'But I was also realizing that time was running out, and hoped that the opportunity I so desperately needed', would soon present itself. But another component in waiting for this opportunity or bringing this opportunity about was the other factor of first needing to get a place in the front row, which would make it easier, and would be the best position for me to be in when the opportunity arose. And up to now I had only got a placed in the front row on four occasions. But I concluded', that I could also try it from the second row if the opportunity came and I was positioned in the right way; but would have to be much quicker and step over people. But later, after scrutinizing the reality of trying to do this from the second row, I realized it would be a much harder thing to accomplish from this position. 'And as these first two days of this last week went by without me even getting the second or third row, I realized even more just how much harder it would be from there? And knew that if I was to have any real chance of pulling this off successfully,' ideally I would have to get a place in the front row. 'And the procedure of gaining a place in the front row was itself a lottery'.

But it was on the Wednesday, during the afternoon Darshan that our line was second to go into the Temple, 'and I did get a place in the front row. And sat there waiting; 'and now knowing that this would be my first real opportunity to strike'. And when later the time came for Sai Baba to enter the Temple I had a clear view of the car entering, and slowly making its way down the aisles, and stopping on two occasions, and letters being taken. And I watched it slowly make its way over to the men's side. Then gradually getting closer to where I was sat. And I knew then that this was the opportunity I'd been waiting for, and if it stopped anywhere near me I would just go for it.

"I had got myself into the right frame of mind; and had my simple statement rehearsed and ready to say it to him. And as the car got closer; eventually turning into the aisle where I was sat; many arms went up outstretched; holding or waving their letters which they hoped Sai Baba would stop and take from them. And by now,' I myself was feeling tense and slightly nervous. And as the car got closer and closer I braced myself and waited, hoping it would stop somewhere in this vicinity, where I would be ready to spring up and approach him. But as it turned into our aisle and I waited for the car to stop and this opportunity to present itself, the car did not stop and just kept going; 'and seeming to go by us quite quickly. And there it was, one more opportunity gone. And this,' was the best opportunity I had had up until now. 'And although feeling very disappointed, I also felt emboldened,' and still determined, and knew that there was still four more days left and of which there would be eight more Darshans: and therefore a possible eight more opportunity's.

And it was two days later on the Friday, when again I got a place in the front row, 'during the morning Darshan. And once more was in pole position, and braced and ready for my next attempt. And as there were now only two days left; and now realising this and feeling more desperate because of it, and knew the chances for these opportunities were running out, I hoped and prayed that this time Sai Baba would stop the car near where I was. And like two days earlier, as I saw the car entering the Temple and then slowly making its way down the aisles, I once again braced myself. And before I knew it, the car had turned into our aisle, and soon made its way over to where I was sat, and I could clearly see Sai Baba, and kept my eyes fixed on him; and hoping this time the car would stop. And then all of a sudden it did stop, and virtually right in front of where I was sat. And he beckoned for some of the many letters to be taken from the people sat around me. And it was then that I saw my chance, and realizing it was now or never, 'quickly sprung up, and managed to move about five feet forward: before the guards stopped me from going any further. "But at that very moment, that I was stopped, I was about three feet from Sai Baba, and he was facing me, and looking me right in the eye: "Seeming almost like he was waiting for me." And the expression on his face seemed to further convey this impression? And at the very moment that this occurred, I said to him, I desperately need an interview from you Baba. And it was then that I was ushered back into my place, and not in a nasty way, but being moved firmly back and being told by one of the guards: 'please sit down sir.' And as I did so, 'the car then moved on.

And although doing what I did had not got me an interview, and there was deep disappointment; I also felt good, and relieved that I had finally done it. And even more pleased that I had got to say what

I had said, and that Sai Baba was looking right at me and heard me, 'and now knew for sure'. And to me, this at this very moment was more symbolic and real than just saying it in a prayer to him. And even more symbolic and special was that I had finally got to ask for, what I had come here for. And even though it had not got me the interview, 'I at least had the peace of mind of knowing, that I had done everything, that I could possibly have done in order to have obtained one. And that I had at least achieved this aspect of my goal, which was to at least speak to him or say something to him; and got to ask for an interview. And I felt that if I would not have done this, or done everything I could to achieve this, I would have left here feeling, that the journey and my time here had been a failure, and a waste of time. But now, regardless of what happened, I could at least now leave, knowing that I had done everything in my power, to have got an interview with Sai Baba. And, for the remaining two days that were left, although I intended to go these remaining Darshans, I also knew there would be no point in trying to approach the car again; 'as I had already said what I needed to say.' And as far as I was concerned,' he was now more than aware of me. And if he intended to grant me an interview he would do so.

But also during this last week, there had also been some real change in my outlook towards my situation; and the burden of it, and what I would do now and in the future. And instead of the bleak depressing idea of having to go back to Britain with the Entity still with me, and all that that entailed. My view of this burden had become more philosophical and more relaxed; and feeling that although it now did not look like I was going to get what I had come here for, I had also decided that I would come back again at some point in the future, and concluded that I would come back one more time. 'And if during that next visit, I also did not get an interview and had to leave in the same circumstances, I felt that there would no longer be any reason to ever come back here again; feeling that there would be no point.' This to me would be confirmation that for some reason it wasn't going to happen. And were reasons for this. And I also decided that whenever this next visit was, it wouldn't be for a year at least. And the reasons for this were various; 'from the journey, to the gruelling schedule. And the heat was another factor, and how uncomfortable it made things. And for these reasons alone the desire to come a back here anytime soon, seemed too much of an ordeal to bear.

But I also thought that coming back here to soon wasn't the right thing to do either. And as well as this, decided that when I got back home, I would just carry on where I left off and still do what I had intended to do anyway. And that was to complete the work on the book; which although now written and brought up to the present time, it now required the finishing touches; comprising of the adding of further notes I had made; and then a final read-through and editing. So although written,' there was still a lot of work to do on it before it was the finished article. And then when this was done the final thing was to type it up. And I also concluded that all of this would probably take me the best part of a year anyway. And I would be in no hurry this time either, and would try to take a more leisurely and relaxed approach to completing the remaining work. And decided, that I would come back to Puttaparthi again during the following September or the next trip after that, which would be in February, and eighteen months later. And whichever one it was I wasn't bothered.

And because of this change in my outlook; there wasn't the despair or desperation I had felt, during the time of leaving, when I had last came here three years before. And, my feelings towards my leaving this time were more philosophical too. And by now with just two days left, I had the feeling and almost certainly felt' that I was not going to get the interview and had now accepted this. And on the Saturday just before going to the afternoon Darshan I wrote another letter to Sai Baba. 'Just a short letter stating to him that I had accepted that it was now highly unlikely, that he was going to grant me an interview this time, or remove the Entity from me this time. And I just simply asked what I needed to do, in order to eventually win his grace, in the form of an interview, or even for him to just remove the Entity by the use of his powers, 'with or without an interview'. And asked for guidance on what I needed to do for the future – what I needed to do – if anything.

And later, as we queued for the afternoon Darshan, and were waiting to go in to the Temple, I had had a very strong feeling that our line was going to be picked first. And I had even pointed to this line, saying to the others that we should sit in this one. And not long after this, my strong feeling was confirmed to be right when our line got first place in the line lottery. And when signalled to go in, up we stood: now assured of a place in the front row. And as we got our places in the Temple and sat there waiting, I also had a strong feeling that my letter would get taken as well. And when the time came and Sai Baba arrived and was driven around the Temple, eventually approaching us; like many others around me I held up my letter. And then the car came to a stop; 'right in front of me, and Sai Baba beckoned for letters to be taken. "And my letter was taken first." And I was pleased about that. But pleased in a very relaxed way, and not overly pleased, or overjoyed. But reflecting and feeling that with my asking him for an interview, in the way I did, and then having this second letter taken from me; that I had done all I could for my situation. And in some ways felt it was my good fortune that I had got to ask him for an interview' and also that he had taken my letters. And that at some point in the future 'these things may well bear fruit. And in the meantime I wasn't worried or anxious anymore. And the remaining two days that were left were positive and uplifting, rather than depressing and desperate, as they had been during the time of leaving on my last visit. And on the Monday and our last day here, after attending the morning Darshan, which would be the last one, we then later gathered with our belongings and waited, for the Taxis to come and pick us up, for our journey to Bangalore. And at 10.a.m. they arrived, and we were soon on our way for the three and a half drive. Then, later arriving there and staying in the same hotel, before boarding the early morning flight to Mumbai; where we changed for the final flight to London.

'The drive and the previous day had been pleasant; and after arriving in Bangalore, I had later gone into the city centre, with another member of the group. Had a look around, the main shopping district, and bought some things. 'But now I was on my way home'. And after managing to spend a good portion of the flight time sleeping, the time had gone by quickly, and we soon landed and were back in London. And after collecting our belongings and saying goodbyes, once again I headed home back to my flat, 'for the next chapter in this continuing scenario'. But as I said, my outlook towards my situation had now changed,' in a way that had made me more philosophical about what I was living with, and still had to continue to live with. And also making the decision, that I would come back one more time, had also helped with how I'd felt. And after getting home, all I really wanted to really do was have a rest for a month or so – before resuming any kind of work on the book. In fact, if anything, I felt like I needed a relaxing holiday somewhere! But I also decided that I would also try to go out more, and try to do some things that would bring me some more enjoyment from life.

But at least one thing that was now 'different, was that I was no longer under any of the pressure' I had been prior to my visit. And this in itself felt good. It felt as though at least some of the burden had been lifted.

Chapter Twelve

Return to Puttaparthi: Part Two

2 Years Later

It was now two years later! And I arrived back in Puttaparthi on the 4th of September 2007. And the year' or eighteen months, that I had originally planned for returning - through unforeseen events, and circumstances, 'had turned into two years'. But, I was back; "and this time on my own."

'And the strange sequence of events,' that had this time brought me back here, once again in 'September,' had began to unfold,' towards the end of the previous year; 'during the November, December, of 2006'.

'I had then been planning to go on the February trip a few months later in the New Year, after having then almost completed the rest of the work on the book. Feeling confident that what was left would be done in time. And like last time, was also planning to travel on the group trip, with the same organizers, and around November had called them to set this in motion. But after having phoned them a few times, and without any reply, on finally making contact with them the first week of December, 'was told that they had decided that the February trip, this time, would be for couples only!'

Obviously very disappointed to hear this, as this trip had been part of my overall plan, and which I had been working towards. And this had thrown it all up in the air.

But, regardless of this sudden spanner in the works, I then had to decide what to do next; and was now feeling quite anxious about the whole thing. And also feeling that maybe I should just stick to my plan anyway and go in February, and go on my own instead. And at this time, just two weeks before Christmas, had decided I would first get Christmas out of the way, and then make a final decision before the first week of January of what I should do. And in the meantime I would also pray to Sai Baba for guidance on what to do, or what was the best thing to do - and was it a good idea to come in February and on my own etc.'

And it was while in Wales, while visiting for Christmas, and in the small caravan I usually stayed in; and just before midnight on Christmas Eve; and while in prayer to Sai Baba and specifically asking these questions. And also, feeling that bit more anxious because of not having had anything that I could say was a clear sign, or indication, or even any feeling of knowing what to do: "That while in prayer and asking these very same questions again; it was during this moment, that I had felt the presence', of a rolling wave of energy around the left side of my head? And then, 'felt it going into my head - in a curved like motion.' 'It was very strong and distinct.' And felt like nothing I had experienced before! And as this occurred,' the word "September" 'boldly appeared in my mind.' This was followed by a pause. And then followed by: 11th September, "interview."

"The initial effect of this was quite stunning!" And although I was more than use to the Entity communicating in a very similar fashion; this whole thing was clearly distinct and very different, 'and had a very different feel to it all'. And although my immediate reaction, was that it was Sai Baba, who had produced this, and was answering me. 'It was soon followed by thoughts of it being mischievous trickery' by the Entity: "My being fully aware of its unique abilities too!" 'Recalling things' like the electrical net etc.' And over the following two hours after this had occurred; I lay in bed, in thought, and pondering over the both possibilities: "Could it be Entity's trickery?" Or was it really Sai Baba!

'This was clearly very clever. And would go hand in hand with the other occurrences and coincidences', relating to the 11th of September and Sai Baba. 'And even though at some point, I had eventually fallen asleep; I had still not reached any conclusion, and was still not sure either way. 'But it was the next day, and which was Christmas day: That something else occurred.' And it was while opening presents' I had been given for Christmas! One of which was off my mother. And which: 'was a Military Cap Badge of the Royal Artillery'. 'And which originally belonged to her father-my grandfather. And who I have no memory of; 'him dying while I was only a few years old'.

He had saw service in the First World War: and this was the regiment he served in. And because he had not had a son himself and my father being a former soldier, he had later given the badge to him.

Sadly, my father had died a few months earlier – in September of 2006. And this is how it ended up being given to me. 'My two other brothers each received a medal from my father's army days', and me getting the badge of the Royal Artillery. "The cancer he had all those years ago had returned in the same place; and this time it could not be operated on! Other things were tried, but after a two year battle it had claimed his life."

The uniquely deeply agonizing thing that bereavement is, the effect, and things that it triggers, hitting what could easily be termed a bereavement nerve; as though touching the very soul itself! I could only sum it up as that. And he wasn't old by today's standards, only 67: And should have been around for at least another ten years.

But he packed a lot into those years. From an impoverished background, as well as serving in the army for a number of years, and many years in the construction industry, also achieved the highest qualification for clerk of works 'class A. And, had also run a Sunday league football team.: "And also served as a County Councillor, winning three election victories as an independent, as well as serving one term as Mayor, and was still a serving Councillor when he died."

In that respect he was someone who got things done; and was held in high regard in the local area. The large turnout at his funeral itself was testament to this. There was not a single seat in the church that was not taken up, and many were left standing.

'There was a lot that was admirable about my father; his intelligence being one of them, and wit another. But also, as my brother said in a speech at his funeral, 'he was also a proper man.' And I could only later conclude after much soul searching and despite my own efforts to get him cured, that for whatever reasons, it seemed that he was meant to go at that time. 'That's what it had seemed like. And so, 'it was with great sadness, that I received this military cap badge.

But, (it was this cap badge) and the occurrences of the previous night while in the caravan,' and my uncertainty as to who was really behind it: 'that had now played a role,' and added another twist to it all of this? And in the form of something I was now' more than familiar with: a coincidence! And which overwhelmingly' I had now regarded, or at least had recognised, as a calling card from Sai Baba.'

"It was a couple of hours' after having opened my Christmas presents, that my mother had been putting all the wrapping paper in a bin bag, and was just about to throw a newspaper in there too, when she asked' if I wanted to have a look at it first. 'My immediate reaction,' was that I didn't want to look at a newspaper on Christmas day. But then said, yeah' leave it there; I might have a look at it later. And it was about an hour later that I had a browse through it. And, as I was looking through it,' I came across a story: "Revolving around a complaint," that a soldier had made regarding injuries, caused by being made to wear ill fitting boots: and apparently was suing the army for compensation.

'The newspaper,' was the Daily Mirror, from the previous day, 24th December. But the story itself was irrelevant. 'But, what was relevant, 'was the regiment this person was attached to: 'which was the Royal Artillery!' "But even of greater significance for me, was that with this article: Was also a 'picture, of the Cap Badge,' of the Royal Artillery!' "The exact same one I had received a couple of hours earlier:" 'The figure of a cannon, and around it an inscription, written in Latin.'

I compared the two, and it was exactly the same? And I also showed it to my mother. And who, although thought it was strange, it would have had no significance. But to me, it soon gradually occurred, 'that this,' may have been the work of Sai Baba.' 'And that maybe it was a further sign', a calling card,' confirming: "that it was him," that had produced the occurrences in the caravan the previous night.' And this was to further confirm this. And doing it in a way, that he knew, I was more than familiar with: "A coincidence! As if to say yes it was me, and here's something else' just to make sure you recognise it.

But even so, I still couldn't be completely sure; and remained cautious. And also feeling that September, was a long way off, and could I really wait that long, if it was him who was instructing me, and was finally going to give me an interview. Even' giving me the date for it; and a date, that had already played a strange role in this whole thing.

But this,' combined with the occurrences: being followed by such a significant coincidence, there was at the same time, also now a gradual growing feeling of that it probably was him. But like I said, was cautious.

But, it was only a week or so after coming back to London. And still no nearer to making a final decision on what to do, when this badge of the Royal Artillery popped up again!

'I had been having doubts about everything. And had been thinking that the occurrence in the caravan, 'probably was', trickery by the Entity! And the whole badge thing, 'was' this time just a coincidence? And I had been feeling quite negative about everything, just generally worn down by all of this shit. The madness of it all; and fed up with what was now seeming more like an ongoing pantomime.

But I had come home one evening and not long after switched on the television. And scanning through some of the channels, had stopped on a documentary that had just started on U.K History Channel. And which was about recent archaeology; as recently as the second world war. And the object of their archaeological investigation, 'was a street of houses,' in Shoreditch, in the East End of London: 'which had been flattened by aerial bombing during the Second World War. And instead of being rebuilt,' was landscaped, and now resembled a park area. And they were digging away and finding various things, and even had elderly people there, who had lived in some of the houses at the time. And who were reminiscing and helping piece together the history and events surrounding the bombing etc. And one of the things they dug up and was cleaned up, and held up and enlarged on the television screen; and obviously in very good condition, and clear to see. Then proceeding to say what it was: Stating, "That this is a Cap Badge of the Royal Artillery", etc etc. And as clear as day, 'there it was again', the exact same one; and almost now seeming as though it was being pushed right in my face? And slightly startled by it; had pondered over the odds of that showing up again, and me coming across it again: so soon; and in such a way!

But it was over the two weeks following this, that I finally decided,' that because of everything that had happened; "I would now give the occurrences, in the caravan, and the following coincidences,' "the benefit of the doubt." And, although I still couldn't be completely sure? Was now feeling to a much larger degree, that 'this all probably was' Sai Baba's doing! 'And this badge' showing up again, was a further signal from him! 'And finally decided I would wait that bit longer and now go in September, and make sure I was there,' for the 11th of September,' now feeling' I had nothing to lose anyway? And during the days after finally making this decision, and overwhelmingly now convinced it had been Sai Baba. I had also mentioned to my social worker what had occurred in the caravan; and said that because of the following coincidences too,' now seeming like a further sign, and her' being aware of my plans, that I had now decided to go in September, and be there for that day. And I had even shown her the cap badge and newspaper story with the picture of the same badge. And I was later asked about it by the Psychiatrist, who she had mentioned it to. And I said what had happened, and a note was made of it.

So here I was again; back in Puttaparthi: "The trip here being much easier this time; "mainly due to there now being a direct flight to Bangalore". And where a Taxi sent by the hotel I had booked was waiting for me and had drove me here. And almost a week had now gone by, now the 10th of September and on the very eve of this date. And now waiting in anticipation for what might or might not occur during the next day. And I had now reached a point where I was now aware, for better or worse: that this was also a date with destiny. Because I had concluded, if I had been tricked, and also left Puttaparthi without getting an interview: "This would also mean that Sai Baba had no intension of giving me what I wanted; and therefore there would be no point in ever coming back again!" As far as I was concerned; this was the end of the rainbow; and that was that. And what I would do then I don't know.

But anyway, everything was set. And in the months in the run up to this I had completed all the final work on the book; and in some ways I was glad that I had had those extra months, because it had taken me much longer than I had expected. But I was now satisfied,' that now, as well as typed up, I had completed every aspect of the rest of this work, and it was now complete, and there was no more I could do to it. And all that now remained,' was an ending.

'So everything was now in place'. And as I write this I am sat in the hotel room I had booked for my two week stay. And have no idea of what awaits me tomorrow! Only that in some way' it will be a date with destiny.

Up until this point and after first arriving, I had quickly reacquainted myself with the Temple. And had quickly got back into the routine, of getting up very early, to ensure a place at the front', or at least in the first few lines. And already had had a close up look of Sai Baba, and he had looked at me a couple of times; and this looked promising. The other good news, was that Sai Baba was now mostly being pushed in a large type of wheelchair; and not always' being driven down the aisles in the car like last time. And this itself seemed to make him more accessible. And up until this point the car had only been used a couple of times. The other change I had noticed was that the wait was now longer in the morning. He was now generally appearing around 9.30 a.m. And because of this, later decided I would attend only one Darshan a day. As doing this twice a day became too much of an ordeal; and besides,' I was here for a specific date anyway.

So, as I awoke on the morning of the 11th of September, and prepared myself and got myself ready, and then set off for the Temple, and prepared for what was now generally a five hour wait; I hoped everything would go to plan' and that finally, I would get what I would come here for: "This was the day!"

Also in my possession was my list of questions,' that I had first written over two years earlier. And which I had by now, added a few more questions to.

'Things got off to a good start. And after initially waiting an hour in the lines outside the Temple our line was finally called, and we were second to go in.' And which ensured I would get a place at the front, and which I did, picking a good spot right in the frontline of the main aisle. And which made me feel really positive, seeming like, that so far everything was going right to plan, and looking like it was all finally going to happen. And with my cushion now firmly in place I sat and patiently waited.

'If you had to leave the Temple for a short while, as long as your cushion was there no one would take your place; this was the general rule. And after being sat there for an hour and a half, now around 6.30 and feeling stiff and in need of something to drink, decided for these reasons I would leave for a short while and come back in an hour. And after having some exercise, and going somewhere for a coffee, I returned around 7.30. And, now aware that although Sai Baba was generally appearing after nine o'clock, I was also informed it could be anytime from eight o'clock onwards. 'So now I would stay put.' And was now in place,' for the final stage of the wait.

The time passed by slowly; and seemed to drag on and on. And as the Temple gradually filled up; all types of thoughts were flowing through my mind; some anxious; and some of them elation like. Anxious that I may have been tricked; and the elation at the thought of finally being picked out, and given the interview, and him removing the Entity: "And then answering all my questions - telling me some important information etc." And maybe he would even materialize something for me, which was something he often did; a ring a chain, or pendant of some kind: 'envisioning a happy ending to it all.'

As time moved along, and finally getting nearer to the time of his arrival. And at this point as I looked at my watch it was now nine o'clock. And now thinking to myself, only half an hour and he would be here. 'But to my surprise,' when the Temple clock neared and then struck 9.30, 'a loud ringing bell repeatedly rung out – something I hadn't heard before.' But was soon made aware of what this was for, as everyone started to leave the Temple. 'It was a signal to say, Sai Baba would not be appearing today, "as simple as that."

I suppose I should have been shocked, or something similar. And there was a slight feeling of that; and also thoughts of feeling that I 'had' been tricked. But still kept in mind,' that there also was still an afternoon Darshan: and that all was not over just yet.

'But on later arriving, for the afternoon Darshan, was told that there would be no afternoon Darshan today, because Sai Baba was attending a sports ceremony and tournament at a sports stadium, 'that had been built by him. But was told I was welcome to go there, 'giving me directions, and it being about a mile away. And which I then went to and waited for Baba to turn up.

'But it was a packed stadium.' And I could quite clearly determine after he showed up, that there was no way' that I was going to get that interview, in this forum or situation; it just wasn't going to happen. And after spending probably three hours there, finally came to the conclusion that my suspicions were in fact, a reality, "and that I had been tricked;" and then left before the rest of the crowd. And the Entity later confirmed this to me, and what I had already long suspected: "That it had produced the occurrences in the caravan; "it, had been responsible for them!"

'And this was also another reason why I hadn't initially been to shocked; because I had remained cautious all along – just in case; and had only been too aware,' that it could have been the work of the Entity too! "But, because of all the coincidences and occurrences around the date of the 11th of September, and distinct nature of the occurrences in the caravan,' had also felt to a large degree that it was more highly likely to have been Sai Baba." 'But I was wrong!' And now I knew the truth. And the uncomfortable truth that it was. And I suppose like any sudden information, concerning very bad news, it doesn't always immediately fully sink in, or take effect, but often,' takes a bit of time' to comprehend or fully register. And later, as I walked along the road back to the hotel,' there was now, more a growing sense and feeling of both, almost shock; and, bewilderment.

But I also tried not to dwell on it too much. And later, during the early evening, went out to a restaurant for a meal. 'But still could not escape from the events of the day;' and the effect and the sense of isolation I was now feeling; 'also now pondering over my fate too.' And also a knowing sense that even though I would be here for seven more days, I was just not going to get that interview!

And over the days following, although I still went to most of the Darshans anyway, just in case I was wrong, still holding on to that glimmer of hope; but once more; like the two previous visits, 'it did not happen.' And once more on the day of my departure I said goodbye to Puttaparthi. And this time knew it would be my last, as I had no intension of coming back.

But even though I had been tricked, and waited for, and had built my trip around this specific September date. 'I had planned' to come here that one last time anyway; regardless of when that was. And regardless of the outcome, I had at least achieved what I had set out to do. 'There was some consolation in the fact that I had at least achieved that'. And regardless of the disappointment, there was also now a sense that I could move on from this, 'I was clear about that!' There was now a sense, of at least that burden being lifted. And not what I would call a new start, but more a closing of a chapter. 'And where I would go from here,' I don't know. In many ways, this felt like the end of the road. "And now all I really wanted to do was get back to Britain."

"And it wasn't just a case of not getting the interview and the repercussions that this brought, but there was also other questions that arose from this, and I now searched for answers to also, i.e. all the coincidences and what or who was behind them? And also there significance! And this also' had had quite an almost stunning like effect too; as there was also a feeling of now not being sure where these had come from either! I certainly needed time to think, and ponder over this whole aspect too – and even the whole thing with Sai Baba.

The journey back to Bangalore', and the flight home went smoothly, with my flight ticket even being upgraded to club class, and was a comfortable and pleasant flight home. Now back in London, my main priority was now just to get the rest of the book completed, and published. And had concluded, that regardless of my own situation not turning out the way I had hoped. "The main reason for my writing this book in the first place, was to reveal the truth behind what for so long has been thought to be a mental illness, a brain disorder, and how this truth was all revealed to me." And this was the real focus of this book. And although I had anticipated,' that if I had got the interview and the Entity removed, I would have almost certainly been told some extra information as well. 'And what I had thought would be something of further significance, of real importance to add to this; 'as well as personal insight regarding my own predicament.' But this was not to be the case! And unfortunately, the book doesn't have the Hollywood ending.

In the weeks following my return I had also given further thought to the whole scenario concerning Sai Baba and the coincidences too? And had looked back on the many occurrences of these; and mainly focusing on the more striking ones. Some of which had sent a shiver down my spine. And the conclusions I had reached is that the coincidences were obviously being produced by something, and for some reason. And what seemed without doubt specifically aimed at me. And I again, couldn't help but think and ponder over the harsh reality of evil power at work in the world; and which this Entity was possibly connected to – or part of. And they may well have been a product of that!

"The other two scenarios that were also a possibility: one of which was something I had briefly read about a couple of years back. 'Calling them, what was termed: synchronicity?' And more or less saying, synchronised occurrences such as these coincidences' were a result of being where you were meant to be, 'at that particular time in your life.' And this was a clear sign of that. And the fact, that so many of these had occurred, and most of which were of a nature, that they were too much of a coincidence, to just simply be' a random coincidence. With this too, and considering my unique situation, this synchronicity element could not be ruled out either.

'And they continued a couple of months after my return'. And two of which were quite brilliant, and were two of the most perfect examples of synchronicity and a coincidence, occurring at the same time simultaneously.

"But the final scenario and of whom I had originally thought was behind them, was, Sai Baba himself! 'Them being the product, of his own extraordinary abilities; of which I could still not completely rule out this possibility either. After all, they had come from somewhere. And had arrived at significant times, and generally overall, seemed connected with my involvement with Sai Baba!

But regardless of the guessing game surrounding there true origin, or intended purpose. I was at least now sure of one thing. And that was that when they now occurred, I would no longer give any special status, or attribute any significance to them. In fact, in truth, I was quite sick of them and glad of this new position; even though some were awe inspiring. And felt' that as I was not entirely sure of their origin, it was now the right thing to do?

And what of Sai Baba! And why didn't he help me?

'Although I had given much thought to the possibility that he either, couldn't get rid of the Entity, or that maybe he wasn't as powerful as was propagated. Or quite simply; the other side of the coin being that there was a good reason, why he couldn't help me in the manner that I wanted – at this point? And he knew that! 'There were still things to do maybe'. "It could even quite possibly be that I was lumbered with this situation until the book was published" Then the mission was completed. This could even be a factor. Maybe then the Entity would just leave. The easiest thing to do would be to discredit or dismiss him in some way, or portray him as bogus or not all he said he was; and blame it on that. This would be the easiest thing, and probably the one that would make me feel that bit better about it all. But the simple fact is, there is too much evidence, overwhelming evidence, that suggests 'he is' genuine. And I as well as millions of others have experienced proof of that. And it would be hypercritical not to acknowledge this reality: 'My own numerous answered prayers being just one facet of this!' But as I said, and regardless of this, or of my own ongoing journey, or where I go from here, this book: The Unwanted Companion, and what it reveals, is the main reason for my writing it. And regardless of my outcome, and what didn't happen, this is neither here nor there. And for this reason, it is on this chapter I will bring the book to a close. 'And of my own situation,' or where this will take me; or what it will evolve into, I don't know?

And so, my own journey continues!

Final Thoughts: Summary

As I have already made clear. The reason for this book is to reveal the truth behind what is thought to be a mental illness. And as I've already mentioned, people being possessed by spirits is nothing new. And there are many books of a spiritual and religious nature that make reference to such things. But my story is unique; and in the sense that it focuses on one type of Entity. A powerful Evil Non Human Spirit Entity, with unique abilities: And the trouble they cause in the world at large. And one that is responsible for what has generally mistakenly been regarded as mental illness, i.e. the worst aspects of what is thought to be schizophrenia. Also unique in the sense that I was put through every aspect of the various multiple abilities of this type of spirit. And shown that what was revealed to me, that there was absolutely no doubt that this was the truth.

During the time span of my writing this book, I had, cut out of newspapers, articles of 16: (and often horrific) random attacks by people unknowingly possessed by these Entities. And all of which had ended in the victim being killed - murdered. And nearly all of which had made the national news; most of which were very high profile: some more than others. And every one of these people had said that voices in their head had either commanded them, or controlled them, or had played a role in their actions. And as with all of them, the article would generally have a headline referring to an attack and murder by a mental patient, or followed by, the person had a history of mental illness, and was diagnosed as suffering from schizophrenia, or a, 'paranoid schizophrenic.'

But these were only the high profile ones, which were reported in the media. And the full scale of this had only come to light when an official government report was released in December 2006: Stating that the actual official number of attacks by mentally ill people in Britain during the last eight years, resulting in a death: 'Was more than 400'. And as well as a percentage of these being random attacks on strangers, a much higher percentage of the attacks were on people who were known to the person. And although I couldn't say for sure, what percentage of these other patients were the possessed kind: 'It wouldn't surprise me if it was all of them!'

But although this is the more horrific side that these Entities inflict on the human race; their general influence in the wider world is as equally disturbing. And which could be summed up at best, as mischievous; and at worst, as evil. As the reason they invade and possess a victim in the first place would testify. It simply being, because they like to live in a body and have an influence in the physical world! 'Albeit: A negative one?' And have none, or little value for things that are generally valued by the human being; and who are generally contemptuous of. And for those of you who sometimes ponder over the question, of are we alone in the universe, is there life on other planets or in other galaxies etc. For the answer to this question is a harsh reality; that we as human beings on earth our not alone in the universe, but live a coexistence with invisible life forms that we cannot see. And some of which are a menace; a real evil that plagues humanity, and is the cause of much misery, trouble, and suffering. And we don't have to worry about being attacked or invaded by aliens from other worlds, as what I have revealed in all its detail, is conclusive enough that we are already under attack by something very alien; as the figures for percentage of population who are diagnosed as suffering from schizophrenia would testify.

What I have revealed to the world is a real horror story for humanity. And it is not my intention, to knock the medical establishments, but, to enlighten them! And the primary focus of my writing this book is to tell the truth, and how I came about this information.

But my revelations do not have only to be a revelation and stop at that. But can also be the first step, in a stepping stone towards the development and building of the necessary machines that I spoke of earlier, to combat their influence. And of which I have no doubt are a real possibility. And I say and emphasize this possibility, in the same sense: that there are many things that we would have once thought to have been impossible to invent, and were thought to be the stuff of science fiction, and which were later invented, and, all of which we now just take for granted. And in this same equation, the invention of this technology is no different. And with the research, development, and a blending or fusion of the right physics and components, this would also make these machines a very real possibility too!

Although, the Entity had never told me, what it, and these other Entities looked like in their visible form. What it did much later say to me: 'is that these Entities', did have something very distinct in their makeup: 'A very distinct feature!' And if the technology is developed, successfully, that this same,' very distinct feature would show up and be seen in the bodies, of all those that have been invaded by one.

And because of what I have revealed in all its detail is not only groundbreaking and the absolute truth, and will give a new understanding to this condition; but like all new discovery's of fundamental truths, at some point it is also accompanied by hard evidence. The initial discovery of the world we inhabit; on being told it was not flat, but in fact was round, and a planet, suspended in space, and which rotates around the sun. This was initially met with disbelief; something that seemed hard to comprehend or believe. But nonetheless, gradually, it was proved beyond any doubt that this was so, and was a fact. And in this same way, and because of what I have revealed to the world, the time will also come when this mantel is taken up, and these machines will be developed. And then the reality of my revelations will be beyond any doubt, and there for all to see; and where then, finally, medicine and science, will touch hands with spirituality.

The End

A Final Note: The Mystery of Joan of Arc

It was no coincidence, that in January 2008 just as I was finishing off the book, that I just happened to come across a film late one night on the BBC, about Joan of Arc. The most recent film that had been made about her life. The first one being made either in the 1950s or 60s, which I vaguely remember seeing when I was a teenager. And was unaware there had been a more recent film made about her.

This latest, the 1999 film Joan of Arc: the Messenger, starring Millo Jovovich, telling the story of Joan of Arc. And from what I gathered was a more detailed and well researched account of her life, and something very much different in its detail from what I could recall from the earlier film. And much of which, must have been taken from transcripts from her trial and what she had said in her defence: As well as from her well documented life. And not a gloss over, or messiah type personification I had remembered in the previous film. In this, a gritty, detailed and well researched account portraying the finer details of the whole story.

As the story unfolds, it is first mentioned that she hears voices. And she is referred to as the maiden from Lorraine who hears voices from God. And she talks of experiencing the reoccurring events, of the strange wind that came; and with it messages.

Her story itself has puzzled and amazed, and also left a question mark as to the true source of her predicament: as to whether or not she was 'divinely guided. But during, and after the watching of this film, my own accumulated knowledge and experiences had led me to conclude 100% categorically; that she had been, no more than a victim of one of these entities, and it was it, through its trickery what had fuelled her self belief and spurred her on. The film portraying and manifesting this coercion aspect excellently. The entity itself also confirming and further pointing this out. And it brought to life perfectly how (the actor Dustin Hoffman 'playing the voices) trick and eventually mislead her to her doom; playing around with her, 'and characterizing exactly, the power and malefic nature of these entities. And I can also conclude, that the mystery and reality of Joan of Arc, was that she simply was a victim of this same thing, 'one of these entities'. These, being all the same and similar variations of these same tricks used and demonstrated on me. Her being tricked into thinking it was God, just as I was initially tricked into thinking the source of my good fortune were angels – seduced by the otherworldly phenomena that I was experiencing. The strange wind she described, exactly the thing I had experienced in varying strengths, to the noble manner of how the voices can be conveyed. Or hallucination type manifestations, like the collage of fast moving images in the sky: something the entity had demonstrated to me many times. "And ultimately its evil trickery! And it is these further finer details like these, seen and brought to life in this film, that are further confirmations', that allow me to say conclusively, that this was her true status. And that my story is also her story; The Unwanted Companion is also her truth, was her reality.